The Possible Present

A volume in the SUNY series in Contemporary Italian Philosophy

Silvia Benso and Brian Schroeder, editors

The Possible Present

Ugo Perone

Translated by Silvia Benso

With Brian Schroeder

With an Introduction by Silvia Benso

SUNY PRESS

Published by
STATE UNIVERSITY OF NEW YORK PRESS
Albany

Printed in the United States of America

For information, contact
State University of New York Press
www.sunypress.edu

Production, Laurie Searl
Marketing, Michael Campochiaro

Library of Congress Cataloging-in-Publication Data

Perone, Ugo.
[Presente possibile. English]
The possible present / Ugo Perone ; translated by Silvia Benso with Brian Schroeder.
p. cm. — (SUNY series in contemporary Italian philosophy)
Includes bibliographical references and index.
ISBN 978-1-4384-3745-3 (hardcover : alk. paper)
ISBN 978-1-4384-3746-0 (pbk : alk. paper)
1. Time. 2. Philosophy. I. Title.
BD638.P4513 2011
115—dc22 2011003714

10 9 8 7 6 5 4 3 2 1

Contents

INTRODUCTION

Struggling with the Angel

Finitude, Time, and Metaphysical Sentiment

Ugo Perone is one of the most lively, productive, and original contemporary Italian philosophers. Born in Turin in 1945 and educated at the University of Turin under the guidance of Italy's greatest hermeneutician, Luigi Pareyson, Perone was schooled in the study of Secrétan, Schiller, Feuerbach, Benjamin, and Descartes in addition to other major philosophers (especially Hegel, Schelling, Kierkegaard, Husserl, Heidegger, and Merleau-Ponty) whose names constellate his numerous books. A continual engagement with theology, most notably that of Barth, Bonhoeffer, and Bultmann, is also integral to Perone's philosophical research, which in recent years has extended to a consideration of poetry (especially Celan) and narrative as areas capable of crucial philosophical contributions. Following years spent in Germany (Munich, Freiburg, Berlin), since 1993 Perone has been Professor of Moral Philosophy at the Università del Piemonte Orientale in Vercelli.

Perone's philosophical activity has never been confined solely to the world of academia; he has always been an ambassador and promoter of culture and education within the wider public sphere. Founder and director of the prestigious Scuola d'Alta Formazione Filosofica (School of Higher Philosophical Education) in Turin, a postdoctoral institution that has seen the presence of famous philosophers such as Jean-Luc Marion, Dieter Henrich, John Searle, Charles Larmore, Agnes Heller, Emanuele Severino, and Jean-Luc Nancy, Perone has also played a fundamental role in the cultural life of Turin and the surrounding region. During most of the 1990s Perone was involved in the administration of his hometown, Turin, as *assessore alla cultura*; in 2001–2003 he was appointed *clara fama* director of the Italian Cultural Institute in Berlin; and since 2008 he has been *assessore alla cultura* for the province of Turin. Through his conjoining of philosophy and politics, Perone strikes a splendid example of the public intellectual that seems to characterize more than anything else the Italian philosophical tradition ever since the Renaissance.

Perone is an original philosopher and an important thinker who in Italy and Germany enjoys a widespread and well-deserved reputation for his theoretical rigor, clarity, and force of argumentation, as well as for the timeliness, amplitude, and suggestiveness of his philosophical positions. An excursion through some major recurrent themes and categories characterizing his thinking will provide ample evidence for his fame.

The threshold—A widespread tendency among contemporary philosophers constitutes and understands philosophy in terms of a *locus minimum* of thought. That is, the metaphysical instances that had informed the Western way of thinking at least up to Nietzsche, but that still continue even in Heidegger, are fragmented to produce, although not exclusively, one of the following three options: (a) specialized branches of philosophy (such as the various forms of applied philosophy and applied ethics) in which the passion for the whole is put aside in favor of sectorialized fragments of it; (b) the value-free and ultimately empty debates of analytic philosophy's analysis of the consistency of concepts, claims, and theories that reduce the richness of existence and experience to a matter of internal coherence; or (c) the weakening of the philosophical horizon in the various hermeneutic positions up to its dissolution in Derrida's deconstruction and, although less so, in Gianni Vattimo's weak thought.

Unlike, although perhaps not against, all this, Perone understands philosophy as a *locus maximum* in which the truth, being, and the very meaning of the existence of the finite subject are at stake. His analysis is existential-phenomenological in its description of the fundamental structures of existence; his horizon is essentially hermeneutical in his constant referring to notions of sense, being, truth, and even the infinite; and his method is substantially dialectical (albeit of a peculiar dialectic, as we shall see) in his searching for the point capable of holding opposites together. Informed by and yet breaking with much recent thought, Perone advances a strong philosophy with equally strong metaphysical ambitions: what is at stake is being, and "against being, which is strong, we are allowed to be strong," he writes (*Nonostante il soggetto* [Despite the Subject], 108). The center of Perone's metaphysical ambitions, however, is no longer (because, after Nietzsche, it can no longer be) God, the absolute, or the infinite, but rather the finite subject that has lived in and upon itself the break that modernity has brought about, that is, the break inflicted by a process of secularization (Nietzsche's death of God) that cannot be denied or easily dismissed. In this sense, although with no explicit admission, Perone's thought complies with the invitation to "remain faithful to the earth" that Nietzsche's Zarathustra passionately advocates.

Perone's philosophy is thus very ambitious; his guiding strategy and his philosophical goal is to maintain the questions of modernity while also accepting the legacy of modernity (that is, the fact that we cannot return to modernity *tout court*). Moreover, his project is ambitious because, in times of fragmentation and deconstruction, his thinking is aimed at designing a full-fledged philosophy, or even a metaphysics, and not simply a way or path of thinking (Heidegger's various *Wege*, *Holzwege*, and *Unterwegs*). Perone's position has in fact the rigorous completeness not of a totalitarian system (to which he is for the most part opposed in his holding on to the fragment against all totalizations), but of a whole capable of keeping together opposite extremes according to the anti-Hegelian (which is also anti-Kierkegaardian) dialectics of the "*neither* this *nor* that," which is also, he maintains, a "*both* this *and* that." What ensues is a "dialectics of dangerous mediation," as he refers to it in his two most recent books, *The Possible Present* and *La verità del sentimento* (The Truth of Sentiments). Perone's style constantly returns to the same topics—the finite, reality, existence—but from different standpoints in order to save the multidimensionality of finite experience by bringing it back to a horizon of sense that needs to be found but also created anew every time. The double negative (neither this nor that) of Perone's peculiar dialectics, which is also a double positive (both this and that), appears in his work as the image, which is also a conceptual category, of the "threshold."

To the threshold of philosophy belong first of all the authors whom Perone engages in a sort of "lateral thinking" that calls such figures into question so as to rehabilitate and make central what, from the canonical perspective of the history of philosophy, has been less relevant in them: God in Feuerbach's atheistic thinking, existence in Anselm's ontological proof, the infinite in Descartes' *cogito*, interruption in Schiller's totality, secularization as a positive legacy in Bonhoeffer's theological thinking, redemption in the instantaneousness of Benjamin's *Jetztzeit*, the self or I as that to which to return after the horrors of a history that we cannot escape in Celan's poetry.

Mutated from Benjamin's *Schwelle*, in its spatial as well as temporal features, Perone's central category of the threshold indicates "not a line but a zone. At the same time, however, this zone that can be recognized only a posteriori, insofar as one has crossed it or has anticipated its crossing in the form of its imagination. Also, it cannot be inhabited but only crossed over. Finally, the one who perceives the threshold simultaneously dilates and deepens it" (*The Possible Present*, 16). In other words, the threshold joins while differentiating and differentiates while joining the here of immanence and the there of the beyond or transcendence, the inside of the familial and the outside of alterity, the finite and the infinite, the before and the after, the origin and the end, the past and the future. Yet the threshold is neither this nor that, and therefore it is also this and that. It is what enables the passage, the move, the transformation, the overturning (again, a dialectical move) of the one into the other—not in the sense that the one becomes the other, but in the sense that every one always has its other, like reality, which always has its masks and shadows.

The threshold ultimately is the present, which one can never possess but in which one always is as the condition in which the finitude of existence unfolds and plays itself out. The threshold is the cipher of the finitude of the subject, and of the passions of the subject for its finitude, a subject who can only live in the present as a historical present pregnant with both the past (the past of memory) and the future (the future offered to us through narrations). In this cipher Perone's entire philosophy congeals: finitude, subjectivity, present, history, memory, and narration constitute the major themes of a philosophy whose meaning, both conceptually and chronologically, is better articulated through a metaphor and the titles of five of his books, namely, *Storia e ontologia* (*History and Ontology*), *Modernità e memoria* (*Modernity and Memory*), *Nonostante il soggetto* (*Despite the Subject*), *The Possible Present*, and *La verità del sentimento* (*The Truth of Sentiment*).

Struggling with the Angel—The metaphor that, by Perone's own admission, has inspired his entire thinking is that of Jacob's struggle with the angel recounted in the biblical book of Genesis. It is the image of the stranger who, in the desert night, interrupts Jacob's solitude and struggles with him in a fight that will end with neither winners nor losers. Only at dawn Jacob will find himself wounded by the angel. The wound will, however, also mean the blessedness of a new name—Israel. Jacob has fought with the angel and has not been killed, and God has fought with the human being and has not won. Rather, the fight by the finite, the fight *for the sake of* the finite has been blessed by the infinite with the recognition and affirmation of the finite through the new name. The struggle thus indicates resistance and tension, it is a resistance that institutes a tension—between the finite and the infinite, the human and the divine, earth and heaven, "the penultimate and the ultimate," to use an expression proper to Dietrich Bonhoeffer, an important author for Perone.

What is relevant in the metaphor is that the benediction, that is, the confirmation of the finite in its finitude, occurs only after the struggle has been engaged, after the finitude has been defended. It is only in the struggle, after the struggle, that finitude constitutes itself as such—the new name means a recognition and an institution. That is to say, the finite can be proclaimed but also valued as such only in its confrontation with the infinite, without which the finite also loses its value and relevance, philosophical as well as existential (as in the perspectives of nihilism and deconstruction, Perone argues). The wound—which is a trace and not presence—on Jacob's thigh is the mark of the infinite. The plenitude of the origin is not with or by the finite, Perone concludes; the struggle does not save the whole, it saves a part and leaves a scar; something to care for, to attend to, to be passionate about. Against the logic of the system, we do not have the whole; we only have the fragment, the finite. Yet, against the logic of dissolution, of nihilism, we do not only have the fragment, the nothing; we have the fragment that has been scarred by the whole; we have something. In this respect, Perone remains loyal to the teaching of Luigi Pareyson—according to whom "truth is interpretation," but there is no interpretation unless it is an interpretation of the truth, as he states in his most famous work, *Verità e interpretazione* (Truth and Interpretation).

Perone's philosophy is this struggle with the angel, which demands that the something that the finite is—history, time, immediacy, the body (still a marginal theme in Perone's works)—be recognized and blessed without one's thereby becoming enslaved to the finite one wishes to protect. The titles of Perone's five major books (but they do not comprise all of his production) mark, philosophically and chronologically, the steps or stages of such a struggle.

History and Ontology—This is the (English) title of Perone's first important book, *Storia e ontologia*, a 1976 collection of five "essays on Bonhoeffer's theology." The book is, however, much more than just an interpretation of Bonhoeffer. It is the, albeit still tentative, proposal of a specific philosophical position that considers history as the site of an ontology; that is to say, in a rather Hegelian and Heideggerian mood, history is the site in which being, the origin, the absolute gives itself. If this is the case, though, being too is traversed by the breaks and interruptions—in theological terms, the process of secularization—that characterize history. Being is not history, and history is not the origin. Yet, in order to grasp the meaning of history and refuse to abandon the historical events to the meaninglessness of their scattered existence, one has to refer to an ontology or to a whole as to the place from which we come but that we no longer are, and perhaps will never be. Being has undergone an interruption, being is fragmented, but the fragments are fragments of being. Both being and its masks are real.

It can no longer be the case, then, of the being of traditional metaphysics, that is, of an originary and unitary being; rather, it is a being that can be retrieved, if it can, only at the end and not at the origin, which is precluded to us. Ontology is thus an a posteriori ontology that is neither presupposed nor granted. It is within history that the search for such an ontology unfolds itself. In this sense, Bonhoeffer's themes of the seriousness of secularization (which is the legacy that modernity leaves us), faithfulness to the earth, the imperative to live life in fullness *etsi Deus non daretur* (as if there were no God), and the multidimensionality of being—all Nietzschean themes—assume an ontological and not simply ethical value. The finite in its multiplicity, what Bonhoeffer names "the penultimate," that to which, in all honesty (*Redlichkeit*), secularization and modernity compel us today, is to be lived, protected, and fought for because it is where being gives itself. The so-called death of God, what in Bonhoeffer's terms appears as "the coming of age" of the world, is not an act of rebellion to God, thus demanding a condemnation; rather, it is the gift of God that makes human beings more human, that is, more finite, but also more capable of love—love of God in the entirety of God's multiple being, but also love of the finite in its multiple aspects.

Modernity and Memory—This is the (English) title of Perone's second important book, *Modernità e memoria*, published in 1987. The themes of the commitment to and engagement with the finite, which as Bonhoeffer's legacy had concluded the previous book, again take center stage. This time, though, the emphasis is on memory because, as the inscription on the cover page declares, "memory is the site where, in a finite manner, the interrupted sense of the whole is deposited." On the background, as the given that cannot be neglected or dismissed, is the interruption that modernity brings about. Modernity is the age of secularization (Feuerbach, Bonhoeffer, Nietzsche), but also of fragmentations and breaks in the continuity of history and the tradition (Benjamin); in other words, it is the category of the *caesura* (a term utilized by Perone) that imposes itself as the most appropriate hermeneutic concept to understand modernity. Two options present themselves to philosophy: to abandon the fragments to themselves in a nihilistic drift, or to try to keep them together within some horizon of meaning. Perone chooses the latter, and there is no doubt that his is a thinking of the whole. But, as one learns by reading (Perone's reading of) Bonhoeffer, the whole is itself fragmented and discontinuous. Memory emerges then as the faculty through which the finite tries to hold on to the infinite. In the fragment that memory remembers, what memory in fact wishes to remember is not simply the fragment but the meaning of the fragment—that is, what is essential.

Following Benjamin, the sign of discontinuity under which the fragments are placed affects memory too. Perone writes, "we cannot, properly speaking, choose to remember. We can only choose to forget" (*Modernità e memoria*, 101), and even the knots on our handkerchiefs, through which, according to an Italian proverb recalled by Perone, we try to remind memory of its memorial task, are powerless as to the object—we forget what we were supposed to remember. It is no longer the case, then, of the "emperor's memory," a memory capable of recollecting the whole thanks to its own totalizing abilities (whether as Hegel's *Geist* in its process of *Er-innerung* or as Plato's recollection of the plane of forms where all souls once have been). Memory is not repetition; rather, it is resistance—resistance for and on behalf of the finite that refuses being either swallowed in the continuity of a system or abandoned to complete meaninglessness and forgetfulness.

Memory does not create such a finite or it would be infinite memory; rather, in a Levinasian move, memory finds such a finite in front of itself as the other, as that which demands not to be forgotten. In this sense, memory is both passive (it receives) and active (it remembers), strong (it possesses the power to respond) and weak (it cannot grant the success of its action), finite (in its powerlessness) and metaphysical (in its faithfulness to a sense of the whole that memory sees present in the fragments for which it resists), tied to the past and projected toward the future. Yet the temporality of memory is the present, because it is in the present that it struggles so that the past may not be forgotten but may instead become a spark of redemption—the happiness that has not been could and should instead be. It is not a matter of having memories, but rather of making oneself memory. The burden then is on subjectivity.

Despite the Subject—The theme of the subject is at the center of Perone's 1995 book, *Nonostante il soggetto*, which was translated into German in 1998. From the outset, Perone is aware of the "regressive aspect" (*Nonostante il soggetto*, 7) of writing a book on the subject today. And yet with courage and honesty that is precisely what he sets out doing, because despite the crisis of subjectivity, of which the various objections to the subject are a stage, the subject is all what we have. Despite the subject, then, we need to move beyond the subject not to abandon subjectivity but to reconfigure it according to new lines of thought. Herein lies the timeliness of Perone's project and one of his anti-Heideggerian, anti-structuralist, and anti-deconstructionist traits.

What we are left with at the end of modernity (which Perone distinguishes from the modern, that is, from the historical period in which the secularization process is carried out and of which postmodernity is nothing except an epilogue), is a weakened, wounded subject, a subject that has undergone a separation from the origin (God or the metaphysical plenitude of the Greek and Judeo-Christian traditions) and is therefore in need of protection, what Perone names "tenderness" toward the finite. It is a finite subject that configures itself through the categories of memory, interruption, desire, patience, attention that is care for the particular, humility, tenderness, and delay that lingers on the finite and thereby prolongs it, as we read in a very agile little booklet, *Le passioni del finito* (The Passions of the Finite), in which Perone meditates on the experience of finitude and the infinite to which the finite inevitably refers. Ultimately, it is a subject that turns itself into memory, that is, makes itself a witness, and a testimony for the other, for a truth, and for an absolute that cannot be said but only revealed in a fact, a gesture, a word.

Of this kind is the subject from which to start so as to accomplish a vision of the whole that neither reduces itself to a system and a totality nor abandons the subject to its own fragmentation and dissolution (which is the other side of the same coin). The subject is thus also the starting point for reconquering a metaphysical standpoint. When we insist on the subject, we in fact discover that the subject is not the foundational site of the edifice of knowledge, as Descartes hoped; rather, the subject is a precarious site constantly manifesting its own inconsistence. Hence, unfathomable depths open up to it: first of all, God and the world, which cannot be reduced to the power of subjectivity, as both Descartes and the various critics of idealism realize.

There is no doubt that Perone's beginning is Cartesian. But his are a different subject and a different metaphysics than the ones offered by premodern and modern thought. Perone's subject and metaphysics have passed through the death of God and have taken it seriously—except they have not been destroyed, but rather strengthened, by such a divine death. And so has the divine itself, because the questions of the truth, being, and the infinite can only be posed meaningfully by a subject that is *there* to interrogate them. There—this means: here and now, in the present.

The Possible Present—We come to the present book, which, published in Italian in 2005, brings together all of Perone's major themes, and therefore constitutes a sort of a *summa* of his philosophy. The book opens with a critique of Heidegger, who has correctly understood how ecstatic temporality constitutes the fundamental dimension of existence, and yet inflected temporality in the direction of the future, that is, of the nonbeing (yet). Against Heidegger and the primacy of the future (the various "to-come" that also characterize Derrida's and Nancy's thinking), Perone vindicates the rights of the present as the proper dimension of finitude and the central dimension of temporality. In its past and future dimensions, time is at stake in the present, because it is within the present that past and future can give rise to a meaningful horizon of life and sense. The present assumes the characters of the threshold of which we spoke earlier; that is, the present is the temporal dimension that enables the difference but also the subsistence and continuity of past and future. The present in fact enables the past to achieve the consistence that allows it to be, and to be even for the future. For this saving activity, an attitude toward the present is needed that does not simply master or repeat the present. Perone indicates hilarity, humility, generosity, and daydreaming as the ways (the virtues of passions) by which one is able to dwell in the present without becoming absorbed by or disappearing in the present.

If the present is threshold, and the present is the dimension of the subject, then the subject, that is, finite existence, is itself threshold, in-between, present that is never stable but always (*à la* Heidegger) on the way to its own becoming present. The I is then fundamentally narration, articulating the discrete nows of the present in the thread and continuity of a story, a tale of the I's becoming and coming to itself.

So it is also with philosophy, which is never the beginning, or at the beginning, but which is rather always after the fact and before the end, as that which tries to give a meaning to what it has encountered in front of itself, namely, reality. In this sense, philosophy as metaphysics is always metaphysics of the finite—what philosophy says is time, the present, the given, the positive. Philosophy says time (the finitude, the present) in order to give it time, to prolong time in the form of a consistent endurance. Philosophy does not simply say the present as is; it also says the present as it could and should possibly be. Philosophy is narration that saves.

The Truth of Sentiment—Perone's philosophy does not stop at its *summa*; rather, it goes on, because there are always more and new fragments of existence the meaning of which needs to be explored and recovered. Reality as object of wonder is what has caused the arousal of the metaphysical feeling par excellence, and hence philosophy. It is to feeling not as sensation, passion, or emotion but as metaphysical sentiment, that is, as sentiment about reality, that Perone devotes his most recent book, *La verità del sentimento* (2008).

Perone argues through a close engagement with Plato and Aristotle, Descartes and Husserl, Kierkegaard, Heidegger, Merleau-Ponty, and Michel Henri, with their accentuations of feelings of doubt, anxiety, nausea, boredom, and so on, that philosophy has progressively dismissed feelings from its field of inquiry and from the realm of knowledge; it has banned them to the private or even irrational sphere, thus losing greater and greater sides of reality. Wonder, from which philosophy begins, is in fact "the thing's urging on the subject," Perone writes. In other words, wonder causes the subject to realize its exposure to the other that the world is and to acknowledge the subject's own relational character. But for such a realization, immediacy is necessary, whereas in the last epigones of thinking about sentiments, "the sentiment, which is born finite (wonder is the immediacy of something), gets to be overturned into being shrine of the infinite"; that is, it becomes perception of the unreachable character of the thing, and immediacy is no longer accessible.

Far from advocating the superiority of sentiment *tout court*, Perone argues for a nonsentimental retrieval of sentiment as metaphysical sentiment, and for a nonirrationalistic critique of the supremacy of reason, since both sentiment and reason, in their intertwining rather than their opposition, are two modalities of the same wonder for reality. "Whereas sentiment is wonder, reflection is the increase in wonder" (*La verità del sentimento*, 37).

Once again, Perone's general philosophical ambition to save all sides of reality—immediacy and mediation, finite and infinite, sentiment and reason—makes itself explicit in the formulation of a "dialectics of mediation" that enables one meaningfully to approach the other by remaining first of all firm (but not fixed) on one's own position because "mediation is not to overcome distance but rather to maintain distance so that proximity may appeal to me" (*La verità del sentimento*, 157). The self and the other in a distance that is the only way of their proximity: Jacob's struggle with the angel, a struggle (a tension) that confirms both without eliminating either, proves to be the guiding metaphor through the metaphysical ambitions of Perone's philosophy.

After the beginning, before the end—To do philosophy is to say the truth, but the truth, as Bonhoeffer has taught (at the beginning) and Perone teaches (before the end), is the protection of reality in its masks and contradictions not only as it is but also as it wishes to be or should be. Not only philosophy but also art, ethics, and, as Perone's own engagement in such a field testifies, politics are all ways in which the finitude of reality in its proximity to the infinite can be said truthfully, that is, protectively. It is to such protection that Perone's philosophy is devoted, after the beginning of such finite and before its end.

Like philosophy, like the metaphysical sentiment of which he writes most recently, in all his books Perone "wants everything and demands everything. . . . [He] is obsessed with the whole. Yet the whole . . . does not have the extension of a totality but rather the intensity of the fragment in which the whole is at stake" (*La verità del sentimento*, 174). It is to the whole that, as a fragment, *The Possible Present* bears witness.

Silvia Benso
Rochester, NY, 2010

Preface

A good preface—Hegel has written some excellent ones—saves the reader, and almost always the critic, from the fatigue of subsequent reading. Properly written at the conclusion of the composition, it condenses the most secret core of the work into a few pages, situates it within the overall production of its author, and perhaps inserts it within the context of the wider contemporary production. This way, if the preface succeeds, the book fails. To everyone's advantage, one could simply exchange prefaces—bottled messages of philosophical intentions.

The opposite venue, that is, immediately entering the *media res* and letting the book speak for itself, forgets that all discourses require a frame—not a summary, which would render the book superfluous, but a situation and a context. Despite all, even books, no matter how complex and articulate, express only a fragment in a path. Understanding their trajectory of meaning, their "to where," and "from where" is decisive. What I am evoking is not a retrieval of historicism, which never gets to the point. Rather, it is the opposite procedure, because the saying, which has said, needs preparation. In recent years, we have paid little attention to the world of finite existence. Perhaps because, being always too far from it—a step back or a step beyond—we have not even noticed that [such a world] was already there. How can we say, however, what we are without at the same time also necessarily taking the step back or beyond that belongs to reflection and philosophy? As Paul Celan writes, "*alles ist weniger, als / es ist, / alles is mehr*" (all things are less than / they are, / all are more).[1] Yet everything, in the form of the too much and the too little, rotates around what is.

Perspectives restore the existing depth to a surface that cannot contain it. To accomplish this, they abandon the immediate rendition of what is, and in a sense falsify its representation. The straight lines of reality become the oblique lines of representation. Observers have neither a simultaneous nor a 360-degree perception. They situate themselves a step outside reality; but they know this, and they denounce it. Their vision therefore succeeds in grasping the piece of world that is being represented. Lamenting this condition would be like bemoaning the optical inversions through which images are produced in our eyes and transmitted to the brain.

This may be precisely what a preface should help us understand—not the subject matter, but rather from where or from what viewpoint the subject matter occurs and, if I may use the expression, on the basis of what expedients it happens.

My most recent book was devoted to subjectivity. The expedient that was enacted there was the viewpoint of the *cogito*, the (already Cartesian) punctual certainty of the I certifying itself against all possibilities of doubt. From there, empowered through such certainty, we could lean out toward what matters the most, namely, the other and the world. One might now be surprised at seeing a book—which is still devoted to finitude—in which the (furthermore obsolete) theme of the ontological proof occupies a central place. With a 180-degree rotation, the intention remains nevertheless the same, as I will try to explain. The same remains also the methodological attitude, namely, the choice for the most arduous and, at least apparently, uneasy way to reach the object of our interest in the conviction that what is gained through the most difficult avenue has greater solidity and more certain proof value.

The ontological proof constitutes the supporting structure of philosophical thinking because it posits the problem of the relation between essence and existence, between what belongs to the nature of a thing and what belongs to its historical manifestation. The ontological proof indicates a point where there is coincidence between essence and existence—one cannot think the one without the other. The proof exhibits the extreme point of such a coincidence, the point where the coincidence is conceived as necessary. In this way, though, it also immediately reveals a specular correspondence with the *cogito*. For the *cogito* too is—at the opposite extreme—a point of coincidence between essence and existence, between the fact that I am and what I am, both being perceived as an indissoluble dimension. Except that here the coincidence is precarious, minimal, merely punctual, and continuously threatened. There instead it is manifest, solar, and majestic. One is a maximal point, whereas the other is a minimal point. Yet, they refer to each other.

In modernity one can begin only with the *cogito*. When explored to its depths, though, this certainty of the beginning point is not (contrary to what Descartes hoped) the foundational site, the ground on which to erect the magnificent temple of knowledge. It is rather a precarious place manifesting its own insufficient consistency. If I want to lean on it, depths that cannot be mastered open up. One can dwell in the *cogito*. It constitutes a safe shelter. Yet, it is a sterile place in which one cannot live. For such a shelter is not sufficient for life. If one leans out beyond the *cogito*, a progressive destabilization opens up and, as it actually happens in Descartes too, it leads to encountering the perfect and full coincidence of essence and existence—God and the ontological proof. This is not in the sense that the *cogito* proves God, but rather in the (certainly non-Anselmian) sense that only God could prove the *cogito*; only the ontological proof is the condition of possibility of the *cogito*, even if, reciprocally, the latter (the *cogito*) is the condition of access to the former (God). The beginning and the ground remain distinct. We could only live in that about which the proof tells us; but we cannot settle and dwell there. Certainty (of the *cogito*) and truth (of God) too present themselves in their noncoincidence.

The paradox lies not only in this condition of laceration. The paradox is also that these fundamental places of philosophy are nothing except expedients. They are fundamental, because they allow us to think and say everything that lies between them. This in-between between that which is inhabitable and that which is sterile is that in which everything is at stake; it is that which truly matters, and it is that which philosophy must try to bring to discourse. The in-between is the place of finitude. The issue is that of incorporating into discourse this place not of the absolute, this place where everything happens yet never in the simultaneous and plastic form of a totality. Bringing such a world of the in-between to discourse means incorporating time into philosophy as its essential theme—time: that is, noncoincidence that is however intertwining [*intreccio*], nonsimultaneity that turns itself into a story.

The threshold of discourse, that which truly wants to be said, situates itself within these two great pillars of thought. Faithful to a radical hermeneutics of finitude, first of all we will interrogate time—which by definition escapes, which is in the form of the having-been and of the not-yet-being, but which also has already been and is not yet simply because it is now, in the present. The task, to which a preface can only point, is to identify time as a phenomenon of threshold, and try to give it consistency. The present, uncertain and unstable, is precisely such a threshold. It is a possible present, both because it is possible for us and because it gives us possibilities. In the present, the voice of time unfolds. There it finds ways, space, and the ease to risk consistency in the form that is proper to temporality and finitude. We will explore time in its *ekstasis* as present, and will try to rethink the existential relations from out of the present for this reason—namely, because one wishes to linger by the present and thereby make it be the in-between in proximity of which it is good to stay. This is even more so at moments when time seems to be taken away from us and forced to be without dignity. One should resist adventurous and desperate escapes forward into the hope for the future or backward into the consolation of the past; one should not abdicate the present, its recovery and expansion.

That which really matters, that which wishes to be said is this in-between, which belongs to philosophy. Even for philosophy, stories, which are neither simply a narrative technique nor a stylistic modality, seem to be the way to articulate the unique and very simple matter that wishes precisely to be said. The I identical to itself, the I as self-referential punctuality, is certainly a firm and stable point. Yet it is also useless because, by itself, it does not extend beyond itself. The I who is full of value and dignity, the I who is person, is a rigorous principle. Yet experience has stepped over it, and has taken it away from us. How can we say the I who wishes to come to itself except that by seconding it and lingering by the stages of its self-searching and its being lost, its being lost and thus having found itself? In the end, how can one say the I in a manner commensurate with the temporal modality that defines it? In this sense, philosophy is always narration, for it tries to say—Plato's dialectic of said and non-said; Levinas's saying and unsaying—in a discursive form that which is very simple but which also can be neither expressed nor properly thought in such a simple manner. Therefore, [philosophy] stretches itself out in passages, articulates itself in arguments, and unfolds itself in considerations.

In order to focus on this simple and difficult matter, in order to say, unsay, and finally say it again, this book works with categories such as threshold, present, and presence, and retrieves the conceptual pair of sense [*senso*] and meaning [*significato*]. Narrating this would be like entering the book. Sober, the preface must stop a moment before all this, but not without recalling that this book has been preceded by essays and contributions that have first tried the issues that find a synthesis here. Deeply revised (destructured and restructured in addition to having been enriched and completed), such essays take the form of this book now, in the present.[2]

CHAPTER ONE

Dramaturgy of Thought

WARNING FOR THE NON-PHILOSOPHER

Philosophers have something in common with children: like them, they play with freedom. Before beginning to play, they love to set up the scene and situate their pieces and any other item in the appropriate place. This too is play, because it is the construction of the *theater*, that is, of the place where the novelty that is going to be played out will come to light.

Children's materials, which are very complex and mysterious as to their most hidden meaning, are very simple in their immediate signification; anyone can easily understand them. If later the imagination fails, however, then one finds oneself riding a broomstick rather than a horse. Conversely, philosophers' materials are complicated and technically loaded. Coming to terms with them without some codified technical skill is difficult, similarly to what happens also in other fields closer to our experience (one could think of the representational codification of a simple geographic map). Such technical skills inexorably lie at the beginning, because they work as the key for what is to come. Despite all my eager attempts, I could not completely avoid all these preliminary remarks and the risk, of which I am well aware, of appearing too technical for the philosophically inexpert, but also (I am afraid) too hasty for the expert. I nevertheless think that documenting, through the evocation of some fundamental scenes, the philosophical scenario from out of which I move is important.

In particular, I think that acquiring these points is significant:

1. In nineteenth-century philosophy and theology, the theme of the present appears as decisive (Barth, Husserl) despite all contrary intentions.

2. Such determinant importance is not accompanied in these authors by an equally sharp recognition of the finite dimension of temporality.
3. In Heidegger, where such recognition is powerfully thematized, the lethal assimilation of the present with the criticized ontology of presence causes disastrous consequences by introducing the primacy of the future, which, far from opening up to the unexpected, has a nihilist coloring and drift.

These few schematic remarks can sustain those who wish to proceed to further details. They could also be enough for those who, accepting these historiographical presuppositions as premise to the discourse, wish to devote themselves to such a discourse and its unfolding. Covering a path without having first indicated from where it sets out and how it differs from its origin would not have seemed fair to me. As is the case for all hermeneutics, the origin is nourished by a dual matrix, philosophical and theological, even if, as is the case for all hermeneutics, what ultimately decides is philosophy, because it is the finite subject who discriminates and dares.

THE ICE SHEET

Through its impossible conditions, theology helps to set thinking into motion. How can one speak of this and that, of the worldly and the otherworldly, the mortal and finite existence, and of the revealed and saving word? Or in any event, how can one say such a word with our words? Within modernity the theologians' impossible task, that is, the task that in obedience they cannot escape, which in faith they know is due but which they never know as guaranteed, is not the application to the supernatural of consolidated philosophical conceptual schemes. It is rather a provocation. Nor is it simply an added productive difficulty, as when the rules of metrics unleash creativity by putting constraints on words. It is much more: it is an impossibility. At the same time, however, it is a command. The dialectics between impossibility and command opens unexpected scenes. The command has an impossible content; yet the command commanding [such content] in turn makes it impossible to live as if the impossible were also inexistent. It is not by chance that Barth himself, and precisely in 1933, titles one of his essays *Theologische Existenz heute*! (Theological Existence Today), and shows that precisely the pure command of theology also originates a manifesto with political value: absolute time is at the origin of historical temporality.[1]

Zwischen den Zeiten (Between the Times)[2] is born within this context and with these meanings. The times between which one is are first of all the absolute times, that is, the worldly eon and the eon of divine time. They are the contestation of the one by the other; they are the ancient Augustinian struggle of the two cities. This conflict between times is also the key to give a name to a widespread and indeed nameless discomfort. It is the discomfort of the human beings who, at the beginning of the new century, no longer recognize themselves because they are no longer the human beings of the past, but neither have they yet found their own consistence. Giving a name to the discomfort means, as Barth teaches, pushing it back into its very ungroundedness; it means contesting what constitutes its strength and asserting that it is not being but neither is it non-being, as it would claim. The old debate on the nature of evil has fluctuated between these two extremes—that according to which evil is non-being [*non-essere*] and that for which it has no reality whatsoever; it is not being [*non è essere*]. What perhaps has not been sufficiently remarked is that evil is a positive claim to non-being; yet it is an ungrounded claim or, more precisely, it is a being that claims to constitute itself in the form of non-being. When pushed back toward itself, when recognized as evil, or religiously as sin, its claim of giving form to being as non-being (this constitutes its strength and even its charm) is unmasked and is restituted to its nature of a weak being that wishes itself to be otherwise, to be non-being. For this reason, once the process of recognition, which actually is tantamount to a purification, is brought to its end evil can even be forgiven, because the positive nothingness has been reduced to nothing, like in a progressive de-coloring that lets only the possible albeit unmasked (and therefore rendered useless) support of evil subsist.

The superficial discomfort of human beings, their sickness onto life [*mal di vivere*], the sickness onto nothingness that pretends to be something and even be the whole, is also called, in the various forms that it assumes and Barth describes, "sin"; therefore, once it is recognized, it is condemned. For this very reason, however, it is also redeemed. As we know, God's No is also God's Yes.

We are always between times, and today more than ever. What is this "between," this thin ice sheet on which one walks precisely between times? It is the present as Barth describes it: the present that is deconstructed, delegitimized, condemned by God's great No; but it is also the present that is constructed, legitimized, redeemed by God's Yes. It is the instant [*attimo*], the almost nothing, as found already in Augustine's *Confessions*. In that sense it is the instant *qua* time that falls and is lost. Or it is the instant of saved time, the *kairos* and the eternal of the encounter between Your history, O God, and mine, the face-to-face, again as in the *Confessions*.

Two absolute scenes: the present is this thin, frail, almost invisible strip of ice separating heaven and earth, almost "the zero-point between two branches of a hyperbola stretching to infinity,"[3] because there we are not only what we are, but also what we are not. "The present time" is called as such "because of that which it obscures and of that to which it bears witness,"[4] and the sufferings of the present time are not "simply characteristic of our life in this world, but actually . . . mark the frontier where this life is dissolved by life eternal. The time in which we live and suffer is the *present time*, the time when glory is made manifest in suffering."[5] The present is the impossible command of eternity, which saves time but, even more, makes it precipitous and precarious, because the instant that ties other instants to itself until it becomes history is rejected. Is it the one or the other, that is, the present instant or the eternal? Or is it the one and the other, that is, the present instant and the eternal? How can human beings cross paths with the eternal without such instants of existence? Does the thin ice sheet separate or unite, oppose or join? It is certainly first of all "crevasse, the polar zone, the desert barrier," cipher of the "distance between God and man"; yet it must be crossed.[6] Transparent and frail, it appears to be both: threshold of opposition and of crossing, without Barth's being bothered by this. Actually, at times he turns it into the "line of intersection between time and eternity, between the present and the future world" that runs "through all history";[7] at other times he enunciates the (equivocal) principle of the "contemporaneousness of all times";[8] and still at other times he asserts the impossibility of "stretching or elongating or developing" from time to eternity.[9] The present stands thus in balance, a step before the abyss, a step before salvation. Yet it remains as the ice sheet that relates the one to the other. It is a difficult, committing present, which could however be also exalting and joyful, as is theology, Barth writes. In any event, it is a present that, if it has a foundation, certainly does not get its own fundament from itself. It is a present that, almost or rather certainly without willing it, in the thematization of the "between times" brings to the foreground the invisible ice sheet of existence, the smooth and fragile terrain of interposition. In sum, when one sets out for oneself the absolute task of speaking of God, even of letting God alone speak, one ends up encountering existence, being unable to elude human beings. Aware of the hermeneutic tangle, Bultmann, who was never understood, kept repeating this to Barth. Thus it happened that for the one who only had eyes for the eternal instant of God's interruption, the present, that is, the place where No turns into Yes, became unexpectedly important, even decisive.

With its prohibition to think finitude and its command to think God's absolute alterity, theology gives us back the thin ice sheet of existence, which can neither be suppressed nor disowned.

TIME UNDER THE MICROSCOPE

As shown especially in the lectures on time which we will consider, for Husserl, a contemporary of Bergson but also—it should not be forgotten—of Barth, the matter is that of following a completely different procedure so as to approach time, and secure the present and its depth.[10] It is not [a case of] the grand scenes of the absolutes, but rather of the minute handling of the watchmaker of consciousness, a working with a magnifying glass, even the patient focusing of the one who works with a microscope.

Husserl's *Lectures* on the phenomenology of internal time consciousness are the attempt to let time surface by suspending its objective dimension through an act of *epoché*. What matters is not time and its duration, which fades away, but rather its lived dimension [*il suo vissuto*], the immanent mode of its lasting and fading. What is the retention, the "comet's tail"[11] of the now that is perceived and then configured as retained now? Or what is the now of the originary impression that is transformed into the now of the retained presentness, the holding of what moves away, the bringing the recent past (that is, the just now) back to the present and the attribution to the just past of an intuitiveness that is proper to the present? What else is recollection, through which I can represent to myself a retained lived anteriority, but a new memorial reduplication of the instant—a reduplication through which the instantaneousness of the instant is pluralized, thereby gaining depth like a probe perforating the smoothness of the sea and scratching the abysses? In the following quotation that has an Augustinian flavor, let us consider the thickening of times of which we have spoken:

> Every tone itself has a temporal extension: within the actual sounding I hear it as a now. With its continued sounding, however, it has an ever new now, and the tone actually preceding is changing into something past. Therefore, I hear at any instant only the actual phase of the tone, and the Objectivity of the whole enduring tone is constituted as an act-continuum which in part is memory, in the smallest punctual part is perception, and in a more extensive part expectation.[12]

What suddenly opens up is an infinitesimal and multiple "between times," which concentrates entirely in the focus of the miniscule present, the perceived now, the now as retention, and the now as recollection, as staying and contraction of a flux, which is bracketed not so much with an exclusionary gesture but rather with a gesture of punctual concentration. The light of the microscope focusing on the instant retains [the instant].[13] Retention occurs, however, within a flux (like the molecule that in focusing comes to the forefront but that continues its own agitation within the flux to which it belongs and from which it is not extrapolated but rather provisionally and visually isolated). The field modification enabled by the microscope again brackets, yet without excluding, the intentional subject that directs its own look toward the instant. It is not the subject that is here regarded, because it rather regards. What is regarded is the act of regarding as an act that produces the reduplication of perception, doubled into retention and recollection (which in the end is retention of the retention, a consciousness of consciousness). It is peculiar how flux and consciousness, which are the two presuppositions of the entire observation, end up disappearing from the horizon thanks precisely to the process of micrologization of perception. They simply disappear from the visual field due to an excess in magnification of the field on which one focuses. On the one side, as we already know, the bracketing of time and more generally of the objective world leads Husserl to say that there is no longer any coincidence between temporal flux as perceived by consciousness and a hypothetical flux of objective time. On the other side, which is the side of consciousness, the warning is no less sharp: "We can no longer speak of a time of the final constitutive consciousness."[14]

Through this path the present, which is instant and retained instantaneousness, acquires a thickness that comes to it from the doubling of recollection and consequently from the interposition of a minimal, micrological, albeit perhaps saving "between times." The retained and recollected retention is the retention whose caducity is redeemed although without denying its character of fading away. Husserl's between times is certainly not Barth's emphatic ice sheet; rather, it is an almost nothing, a trifle, which cannot be perceived outside the work of observation carried out by phenomenological focusing. Nevertheless, this light thickening produced at the margin between retention and recollection is a recollection that is endowed with consistence; it is a stayed retention: the thin thread enabling flux to unfold and reveal a direction.

Because it is endowed with thickness and depth, the plural projects a shadow of itself that is similar to the growth of flux in consciousness. Depth is not only a vertical but also a horizontal thickening, and the internal time of consciousness extends in all directions. Recollection is, in fact, not only memorial retention of retention; it is also recollection of recollection that projects an increasingly wider but also fleeting cone (up to imperceptibility). One should think of Husserl's diagram, commented on by Merleau-Ponty and Ricoeur, and of its complexity, which is appropriate to the willingness to restore depth to the simple plane. Here depth is not only the volume and thickness of time, but also its fading away, which is, however, not a going away. The frequent acoustic metaphor with which Husserl accompanies his lectures gives a good idea of this wave; once it has reached its exact depth—that is, the "tone" that intentionally belongs to it—the wave expands on all sides through the play of retention and recollection exactly like a sound, up to its own dispersion, which perfectly coincides with its maximum expansion.

Here, too, the between times that Husserl construes is peculiar. As already seen in Barth, the Yes and the No, salvation and negation, end up intersecting and in the end coincide perfectly. The present exhibits, contains, shows, stays, and recollects them both. One should therefore turn to the present.

The thickened line of this present is difficult to retain and withhold. The scenes outside the visual field become present again as soon as one turns the microscope off. It is not true that here the *epoché* is an act of intentional exclusion; it is rather the consequence of an intentional inclusion that intersects and pierces through, and then works on instantaneousness. Bergson and Merleau-Ponty are haunting. Consciousness, which Bergson turns into an object and not only a subject, encompasses time and unrolls it as its own internal object. Where is here the present, which is by now the eternal albeit in the form of the eternal and coalescent becoming of evolution? Where is the precarious between times of the present, which is, conversely, entirely contemporaneous to itself, in its own inclusive and implosive retention (which is suffocated, up to explosion) in a ball—an image that is also a regression to archetypes of a petit-bourgeois childhood? Does not the boiling flux, which is removed and denied by Husserl out of prudence and is instead uncovered by Merleau-Ponty, risk opening up a boiling superabundance of the flesh (the new sayable name of the divine around us), which is like a perennial source of youth, a yes-saying to the flux that liberates from all necessity, so characteristic of finite human beings, of delimiting and marking their own territory, of speaking not of the whole but of something?

In Husserl such a something is certainly very close to a trifle. Only the extreme seriousness of a rigorous and even a bit pedantic scientific apparatus can keep at bay the impending impression that, in the end, [such a something] is so miniscule, so only micrologically visible that it is not much more than an almost nothing.

FINITE TIME

Heidegger's great merit, or at least grandiose intention, lies in the vindication of the finite character of temporality: "*Die ursprüngliche Zeit ist endlich*" (Primordial time is finite).[15] In doing this, Heidegger aims exactly where the other authors had failed. Forever, for us, he ties time to finitude, and consigns it to us as our question, as the question in which what is at stake is ourselves.

The complex strategy he unfolds, on which I will not dwell, contains two points on which I would like instead to pause. The first is the criticism of *Vorhandenheit*; the second, strictly related and even consequent to it, is the primacy of the future.

What is important to emphasize above all are the equivocations produced at the level of the interpretation of temporality because of Heidegger's fateful intertwining of critique of the ontology of presence and determination of the temporal *ekstases.* Heidegger himself admits to this when he writes:

> the ancient interpretation of the being of beings is oriented toward the "world" or "nature" in the broadest sense and . . . indeed gains its understanding of being from "time." The outward evidence of this . . . is the determination of the meaning of being as *parousia* or *ousia*, which ontologically and temporally means "presence" [*Anwesenheit*]. Beings are grasped in their being as "presence"; that is to say, they are understood with regard to a definite mode of time, the *present* [*Gegenwart*].[16]

It follows that a conception of time as present time, which reduces it exactly to something simply available, is made to correspond to being as *vorhanden*—that is, as attainable, at hand, as is suggested by the turgid consistency of the German [language] even in its abstract formulations.[17] The present, unlike the past and the future, is then the time that one has (in the sense of having at hand). What can one have in time though? Nothing more and nothing less than what it is right [*lecito*] to attain—in other words, a thing. Yet the present time, which one has, is a reserve of objects [*oggettuale*], something in-between banal everydayness and imperturbable eternity. The present that one has [*il presente "avuto"*], the present that is a property to attain, reminds one of the bourgeois husband who, in the Kierkegaardian living room, hears the strokes of eternity sound in unison with the tolls of the pendulum. We should beware, though: such time is objectified and rendered eternal; it is the presence of the present, that is, the present made to be presence.

To save being as event one must then destroy the ontology of being as presence; to save the temporality of finitude one must deny the present as available presence. The intense activity that Heidegger sets into motion is simultaneously complex and subtle, because he intervenes with a sharp correction on each of the points that describe inauthentic temporality. The time one can have is, as we said, the simply available time. Yet the knot to which the expression "to have time" refers is essential also for originary temporality. One should not abandon such a dimension, but rather reinterpret it in light of the analogy (which, Heidegger warns, is not a perfect analogy) with what one means when one says "to have *Angst*." Neither time nor *Angst* is the core of such expressions. What matters in them is rather the having, which is here understood, in opposition to what has been said before, as that which is "to me" [*a me*], which concerns me, which characterizes me, and not as that which I have a right to attain. This modality of having, for which the reference is neither to possession nor to the object, but rather to the subject and to being "to me," is the fundamental element.

The analogy with *Angst* serves to relinquish the attention on time as an object that is had and to bring it instead on that which properly constitutes the temporalization of time, on that which Heidegger calls the temporalization of dwelling, and on the time that is "to me." If one misses the dimension of having time as the condition for which time is "to me," and if one transforms time into something simply present (*vorhanden*), then certainly one completely loses authentic time.

In this second sense, the time that one has (and that will be revealed as the time that one is as opposed to the time that one has in the sense of possession and availability, which is not authentic temporality) is no longer present time, no longer a "now"; it is not even properly datable, nor does it have the character of publicness (of being shared). Nor is it the time for something or for a determined project (as is disclosed in the analogy with *Angst*, which is anguish over nothingness). It is rather the time that, in an anticipating manner, actualizes the time that is not yet—the future.

We are here in front of the second element we mentioned, namely, the anticipation of the future that assumes a sharply nihilistic tone because the future is a form of presencing [*presentificazione*] of an impossible presence. The future is anticipation not of something—which would be present—but rather of an absence. In such a way presencing does not appear as a halved reintroduction of presence, as some sort of merely virtual presence, which would nevertheless remain a form of presence. Rather, in perfect correspondence with what was said about having time, it is a way of destroying the character of being-an-object [*oggettualità*] of the object. Presencing is Dasein's anticipating attitude, the same way in which Dasein's having-time attests to its temporal character.

Heidegger's attempt goes in the direction of rooting temporality in the finitude of Dasein. To avoid all risks of ontological essentialization, he chooses the primacy of the purely anticipatory modality of the future. "*Zeitlichkeit zeitig sich ursprünglich aus der Zukunft*: Temporality temporalizes itself primordially out of the future."[18] The consequence is that in order to escape the inauthentic temporality of the now, which is reflected in the ontology of *Vorhandenheit*, one consigns oneself to the primacy of the not-yet, of the instant [*Augenblick*] as manifestation of the time that passes [*vergeht*]. In the form of authenticity all this is possible only because the caducity of time is grasped through the anticipating knowledge of one's own death, which one fleetingly experiences in the instant.[19] The extreme attempt at situating oneself on the side of finite temporality is accompanied by the anticipating consummation of time. In this context Dasein itself appears as inexorably inclined toward the direction of the not, so much so that it can find its authenticity only through the transcendental anticipation of non-being, that is, of death. What emerges is a temporality inexorably attracted toward the ending of finitude, and unable to grasp the initialness [*inizialità*] of time. What also derives is a life that is thought within the transcendental horizon of death[20] and an ontology that is marked by an immanent nihilistic outcome.

Heidegger's approach displaces classical ontological conceptions, in which the consistence of being in fact appears as void of all temporal markers and therefore turns out to be nonaccessible to finite beings. The price paid for such a result is, however, not inferior to the gain that is achieved. Heidegger's perspective is fascinating and efficacious in its wolf-like wandering around its prey (as Levinas suggests to me),[21] and in its delineating the field of the phenomenological question. Yet it appears as drastically poor and lacking when it abandons the strategy of delaying and indicates in the direction of an answer. The never filled incompleteness of *Being and Time* appears as the greatest gift left by the philosopher to his posterity.[22] Nevertheless, retrieving Ricoeur's expression in reference to Hegel, this "construction site" too must be abandoned; letting go of Heidegger appears as a necessary step. Heidegger does not succeed with his proposed endeavor—namely, to rethink Aristotle phenomenologically (against Hegel and his immense claim to end the *Encyclopaedia* with Aristotle's *Metaphysics*). Yet the theme that he leaves as a legacy (and this is truly in tuning with Hegel) is not less timely just because of [the failure].

PRESENT AND PRESENCE

No matter what, credit should be given to Heidegger for having brought the intertwining of ontology and temporality back into the agenda. For many other authors (but certainly not for Bergson when at his best), the theme of time was a precious, happy, and delimited island. On the contrary, in Heidegger the radical vindication of the finitude of time is coupled with a critique and a radicalization of the ontology of presence. Retracing the history of nineteenth-century theology in parallel with the vicissitudes of ontology that culminate in Heidegger's critique would not be difficult. The century we have just left behind has been tragically crossed by the conflict between a (mostly Catholic) theology of presence (that relies on a metaphysics of substance) and a theology of the event, the extreme of which is already Barth (with a position that no one of his followers—one could think of Bonhoeffer's attempt in *Act and Being*—could ever maintain in its radicalism). The success of the radical position (of a Barthian kind), better suited for a critical use, is guaranteed in the very act in which the contrast between the two positions gets configured as an alternative. Yet the success that is thus achieved (how could one not agree with Barth in his prophetic unilateralness when one is confronted with the impasses and compromises of the always uncertain theologies of mediation?) leaves intact the question of understanding how the theological event (the unrelated irruption of absolute alterity) may not end up producing, against itself and its own intentions, a secularized and ultimately nihilistic outcome like Heidegger's event. The latter in fact powerfully reinaugurates the sacred (hence the frequent theological likings for Heidegger) yet at the price of having completely dried up God's transcendence. As Pareyson's Naples lecture confirms, Heidegger can be better understood "when one considers his fundamentally non-Christian or anti-Christian inspiration" and understands his nostalgia for Greekness as "a firm relinquishment of Christianity."[23] Heidegger's attitude toward Christianity is in perfect correspondence with what Benjamin says about himself: "My thinking is related to theology as blotting pad is related to ink. It is saturated with it. Were one to go by the blotter, however, nothing of what is written would remain."[24] This operation of absorption/cancellation is the extreme result to which being, which has been transformed into event and interpreted in the form of future event, yields. It is not by chance that Heidegger's God is still to come, that it is a new and future God. Despite its different rooting, the theology of the pure event also faces similar issues and is progressively induced, as the early Moltmann has shown, to accentuate the dimension of the future, the eschatological feature of the coming God, so as to give consistence and understandability to God's alterity. In this manner the event that does not occur swallows the events that do indeed happen in an obscurity in which all cows are the same. The event-like character of the wholly other succeeds against all forms of presence. How can one protect, however, thought and existence from such a terrible victory?[25]

As is already the case for Heidegger, Barth's construction site too does not appear to be further viable despite its many good points. This is so in the sense that one cannot remain where Barth found that he himself had to be. In order for Barth's correct claims to continue to hold, one must first go against him even before going beyond him, as Dietrich Bonhoeffer already did in a form of extreme faithfulness to his master. Barth's construction site must be abandoned because the place of theology is not exhausted in such a space. Philosophy and theology thus find themselves in front of a parallel vicissitude, which has given rise to an analogous and analogously unusable alternative. The urgent need common to the two disciplines on the one side confirms their unavoidable intertwining, and on the other contains some opening of positivity. If one were to find it, the appropriate way out could indifferently come from either of the two disciplines; yet it would presumably hold true for both, and would help both of them to set back into motion a way of thinking that the stiffening of the oppositions has by now frozen.

CHAPTER TWO

The Present as Threshold

Retracing the surface of the world with one's gaze, drawing its profiles, following its recesses, and rendering nature human and available. Without threats. The German landscape, however, has no surface, writes Lenz. In the far North everything, even the sun, takes up a severity that impresses and alienates. There, it is as if everything were depth: without surface, without levity. A damned, threatening depth. "This countryside of yours doesn't know what fun is, not even on a day like this. It's always so solemn, even when the sun is shining. There is this austerity hanging over everything. . . . Even at noon it's all uncanny. Sometimes I haven't been able to help thinking that this countryside here has no surface, only . . . I don't know how to put it: the depth of it—it's got nothing but this baleful depth, and everything that's down there is a threat to you."[1]

In Nolde too (and the nearing is not casual, because Lenz's *Deutschstunde* is precisely about him) the colors spread out as if to enclose depth into a desperate expressionism. The now of time seems still, in a zenithal light that produces no shadows, except for the shadow that things harbor within themselves and that crushes everything in the ultimately threatening presence of the present.

Should we not render such time (and such space, sky, color, and unbearable depth) habitable? Should we not learn to look into such implacable halt in the atmosphere? Should we not unhinge it from the inexorable eternity and threatening scream of what appears to be already past, even when one stops the image of it in front of one's eyes? Should we not recover, albeit in depth, also the surface—and perhaps once again recover the depth from the surface? To say it again with Lenz, "'So you regard it as something terrible, Teo?' 'What I mean is simply there's so much that's human in surface.' 'I know what you mean, Teo, but if it's really like that, oughtn't we to try to make it habitable—this countryside, I mean?'"[2]

Let us return to the present. Let us make it possible, let us make it habitable.

THE PRESENT IS NOT PRESENCE BUT DIVIDE, CONDITION OF TEMPORALITY

We must go back before Heidegger and his curse that reduces the present to a form of presence. We must cautiously start again with those fragments of present that have been snatched away from conceptual and temporal consummation. It is true that we only know the present as (just) passed or as not (yet) occurred. It seems then that the name that best fits the present should be "past" or "future." Yet this is so only because the present—which is time—passes and does not let itself be grasped except than as that which is not (any longer) or that which has not (yet) occurred. For this reason, one takes shelter in the past or in the future, because they appear as secure modalities of time, removed from the flux to which time itself conversely exposes us. In sum, we do not dwell in the present because it escapes us. Because we do not want to lose temporality, we turn to the present as past or to the present as future. In this manner, however, we have lost not only the present but also temporality, which has been petrified in the form of the past or has been rendered volatile in the indeterminacy of the future.

The present is not eternal, otherwise it would have neither extension nor would it appear as threatened, subject to being lost. Nor is the present the now, that is, pure subsistence of what is determined in common, unless because there is a precarious, nongranted belonging to the same present that is common to more than one subject. The present is not the instant either, the pure interruption that rises to hypostasis, the dominating self-insinuation of nothingness (which is invasive) within the horizon of being, which is precarious and needs protection precisely because it is interwoven with time.

We should not let ourselves be paralyzed as if by a Medusa [*medusizzare*] by the incessant flowing of the present. We should not look only at its temporal, albeit essential, structure. The specific difference of the present, the one that distinguishes it from temporality in general, is not its mere passing but rather the divide [*discrimine*], that is, its being that which separates the past from the future (and constitutes them as such). The present is that which discriminates, the passing that withholds, the hand that closes up so as to support itself and others [*per reggere e reggersi*]. Such a divide can be so short, unexpected, and extraneous that it holds only for an instant, which is nevertheless decisive because it divides times and ploughs through history by virtue of interruptions. [The divide] also designs the now to which we belong reciprocally and that crosses shared paths with us. Somehow [the divide] can also be the eternal, and it can constitute an image of it in its having secured forever something that could no longer be and that was drawn to being and saved through it.

[The divide] is all this in the form of temporality—of remaining and fading. It is existing temporality; it is the existence of temporality. It is temporality that risks itself in time, and not Heidegger's withdrawing of time into temporality. In any event, the present that discriminates is not presence. It is never at one's disposal, nor is it simply there. Nor is it the fallacious image of an always already secured presence (assuming that there has ever been a metaphysics so simplified as to think something of this kind; in any case, gaining eternity too takes such a great effort).

Concentration on the present is not the latest variation of a time that serves only the definition of an attitude toward the world. Such happens when the emphasis on the past translates into conservation, the emphasis on the future into will to change, and the emphasis on the present into praise of temporariness or, in a peculiar alternative, into will to institutionalization. This is not the interpretation of time I am looking for. In any *ekstasis*, desire is indeed at work—desire to preserve, to change, or to build a tent or a house. This is time as desire, though; time that is mine and that I project onto the world, the rhythm of existence that wishes to bring the world home. Ultimately, it is always nostalgia, albeit in various modulations.

More essentially, time is something else: it is that which I do not have. Not in Heidegger's manner, for which having time eludes presence through the *escamotage* of the primacy of the future; in the end, the time I have is then the time I do not have because it is future. My perspective is here the reversal: Time is essentially and originally that which I do not have. Such not-having that traverses and marks me is nevertheless the unbalance on which I build my life daily. The present is the threshold where the not-having of time (subjective genitive) intersects existence. It is the fleeting intersecting of an encounter. Life is the always recommencing attempt at expanding such a small opening and controlling time by seconding it [*assecondandolo*] without merely suffering from its offenses. The time that one does not have is then [time] to be had. All this, in truth, eliminates presence, because in the form of objectification presence confronts one with the contradictoriness of having and not-having. Here having is instead a seconding, where activity and passivity are equally legitimized, and moreover, where the subject of having is from the beginning confronted with the alterity of something that escapes not because it is other (intolerable positivism of transcendence). Rather, [the subject is] confronted with the alterity of a time that is essentially fleetingness and that therefore confronts the subject with alterity (from which it would nevertheless like to protect itself).

The present that is not presence, that is, the present that I do not have but on the basis of which I have time, does not divide the world into two, like a summit from which opposite sides depart. The present instead builds the world because it makes [the world] exit a vague flowing by anchoring it to a point, to a subject, and to a perspective from out of which the world appears as that which surrounds me. Then [the present] cannot be a simple limit, that is, [it cannot be] the extreme line either of a contact defining itself through separation or of a being whose consistency is given through exclusion or of an encounter stiffening differences. Nor is it properly a barrier, because the barrier is simultaneously insurmountable and independent from me. The divide is instead mobile, and a constitutive trait of the deciding subject. Nor is [the divide] the irruption of the absolute that tangentially encounters the infinite, the *eschaton* that becomes time, because the divide does not discriminate absolutely but rather chooses on the basis of finitude, reorients finitude, and arranges it within the order of time.

This divide that is neither limit nor barrier or irruption of the transcendent is the threshold; finite matter that is handled by finite existence in order to orient itself within the temporality of the world.[3]

THE FEATURES OF THE THRESHOLD

A series of features that appear to be meaningful belong to the threshold. As Benjamin remarks, [the threshold] is not a line, but a zone. At the same time, however, this zone can be recognized only a posteriori, insofar as one has crossed it or has anticipated its crossing in the form of imagination. Also, it cannot be inhabited, but only crossed over. Finally, the one who perceives the threshold simultaneously dilates and deepens it.

The first aspect highlights the fluid character of such a concept. The threshold is not a line, except an imaginary one. There are numerous traits, which Benjamin efficaciously describes, that document such an aspect.[4] The threshold is the space of a peculiar co-presence of disquieting and familiar, like the doorbell, which discloses but also protects the intimacy of the house. "The despotic terror of the hand bell, the terror that reigns throughout the apartment, derives its force no less from the magic of the threshold."[5] Or the threshold has an almost magical power. The triumphal arch does not celebrate victory; rather, in triumph, it institutes the victor as such. Before passing through the triumphal arch, the winner is someone who has won a victory outside the borders of the homeland. Now, however, he becomes someone who is celebrated as victor within the borders of the homeland. The victory, which is external, is no longer enough. As if by magic, the victor becomes the triumphant: "The Roman victory arch makes the returning general a conquering hero."[6]

Yet the triumph has annihilated the consistency of the victory, despite the fact that it has generated a new condition. Furthermore, once it has been crossed, the threshold of writing renders available to the child a very powerful instrument of communication and magic—to create and send words; yet this annihilates the mysterious and undecipherable world of signs and their arcane power, of which only the adults seemed to have the key. The children who cross over such a threshold do not even realize that they are doing it. One day the illiterate who with difficulty paused on signs and spelled them finds himself or herself a reader.

> The longing which the reading box arouses in me proves how thoroughly bound up it was with my childhood. Indeed, what I seek in it is just that: my entire childhood, as concentrated in the movement (*Griff*) by which my hand slid the letters into the groove, where they would be arranged to form words. My hand can still dream of this movement, but it can no longer awaken so as actually to perform it. By the same token, I can dream of the way I once learned to walk. But that does not help. I now know how to walk; there is no more learning to walk.[7]

Precisely for this reason, the threshold, which belongs to the order of the symbolic, can never be perceived in the form of a presence. It stands in front of me as an insurmountable obstacle with which I confront myself. Or, it is the already crossed-over border, which with similar irrevocability stands behind me, having been overcome or lost once and forever. The threshold is not; rather, it has always already been; like the front, with which I am always confronted, or the crossed and lost stage, which no longer belongs to me. In any case, the perception of the threshold is always the perception of a temporality in which I am not. In this sense and from this perspective, despite the spatial origin of the metaphor, [the threshold] seems capable of expressing an essential element of time: its structural feature of fleetingness, its essential never-being-mine, and the cipher of negativity that belongs to it.

For this reason the threshold cannot be inhabited. In fact, it never belongs to anyone, nor is it anyone's; rather, it is only for someone. The threshold is the not-mine that nevertheless is for me. It is the place and time of the crossing; the path that is proposed to me, the memory of what has been crossed, the waiting for a pass. It is not, however, the place and time of dwelling. The threshold turns [things] upside down because it transforms the over-here into an over-there, the inside into the outside, and the mine into the other's. It affects the I that crosses it, which becomes alienated. The threshold installs itself in the I, and not the I in the threshold. The time that the I covers indeed traverses the I, and turns it upside down.

Finally this crossing, which I always perceive properly only insofar as it has happened, is never exhausted. The threshold that the I perceives is not a border that the I has set up for itself; it is rather a barrier that the I has come across and that has induced [the I] to a reversal. The threshold that is behind me is never behind my back. The threshold that has been overcome is not eliminated; rather, it is withheld in memory. The threshold is even the protrusion of the memory of that to which we no longer belong. The threshold is severe because it destabilizes; it is severe because it never lets one go. It protrudes into the I, who thought itself able to dominate it, and brings to the I the not-having-time, which is the time that one has in the present.

SPATIALITY OF THE THRESHOLD: OUTSIDE/INSIDE

Two are the possible universes of the threshold, that of space and that of time.

Let us begin with space. The threshold is first of all the threshold of the home. It is that which traces the space of intimacy—the inside, in front of which is an outside. There would be no inside if there were not also an outside. Oppositions and conflicts thus situate themselves in the most intimate place, that is, in the originary [place] of dwelling within which being too configures itself. Nevertheless, conflicts are also that which bind together because while it marks the limit, the threshold also makes the reversal possible in the form of a passage from inside to outside and from outside to inside. The threshold defines a world; it is the exit from a chaos where there is no conflict but only indistinctness and indeterminacy. For this reason, the archetype of the home is the universal of protection; it is a way of saying space while making it exit the pure neuter of measuring. Everyone has a home or, no matter how poor [they are], a corner for themselves.

As the threshold is not exteriority, so too it is not the home. The threshold, in which one does not dwell nor is, nevertheless enables dwelling and staying. When the dwelling is understood on the basis of an awareness of the threshold, the I's interrogation regarding its own being arises. The answer extends the I beyond its own dwelling toward the exteriority that the I has simply suspended while inhabiting the home and that the mobile trace of the threshold protects. The I's own very self-understanding varies according to the modalities through which the threshold of the crossing between outside and inside, between corner and home has constituted itself. To put it differently: the threshold, on which one is, constitutes the place of dwelling. Dwelling, however, which is recognized through the difference from exteriority that is made possible by the threshold, makes the question arise concerning the "who am I?" of the one who dwells. The answer to this question is in turn strictly related to how, through the threshold, the relation between outside and inside is understood (as conflict and contraposition, as possibility and extension, and so on).

The modality outside/inside does not only have to do with the dwelling. It also concerns corporality, and more precisely the relation between body and mind. Despite a very strong accentuation of the importance of the corporeal, the materialistic thesis no longer finds supporters in our culture. In general, one presumes in fact a spirituality of the body, at least in the form of its completely human specificity. The body that is at the center of interest is in all cases the body of and for the human beings, that which is not only *Körper*, mere physicality, but also *Leib*, body belonging to me. What truly matters today are the dimensions of the threshold between the outside and the inside, the spiritual and the bodily, rather than the mere opposition between body and spirit leading to a Manichean dualism. Once again the procedure is similar to the one we have described in the preceding: the body that is recognized as always mine introduces its alterity from other bodies, because the identity of the body as my body makes me perceive the dimension of the threshold and the precarious identity/relation, belonging/alienation that constitute me. Once again, different understandings of one's own very identity emerge according to the way in which such a phenomenon of the threshold is played out (as conflict, contraposition, extraneousness, will to power, potential violence, or instead as relation of recognition and solidarity in a finite physicality).[8]

As a confirmation of this, we can perhaps here indicate two examples, which are very different from each other yet convergent. The morbid exhibition of sexuality is certainly also an attempt, in the form of repetition, at mastering its meaning and the spirit animating it. The child takes the toy apart to the point of reducing it to pieces in the search for what is inside, and thus reveals a metaphysical soul; likewise, sexual emphasis and reiteration wish to come to the discovery of a sort of ultimate nature. The child's game already denounces the illusoriness of the procedure, because inside the toy there is nothing and the metaphysical anxiety turns into delusion. This is so not because inside there is really nothing, but rather because what is there is neither inside nor outside the toy; in other words, the threshold of the imaginary that the toy constantly touches, nourishing and exalting the fragile subjectivity that plays, is its spirit. Similarly for sex, where what is properly at stake is the value of the threshold (and not of transgression, unless as constantly withheld transgression). Here, too, the subject properly exposes itself and lets itself be dragged into something greater than itself.

Of the same kind, albeit on the grounds of completely different conditions, is the (often mediatic but at times also judiciary and social) game of the moment of truth, where the word that is said, repeated, thrown into one's face, or exhibited wishes to strip us naked. We often say that we are searching for the truth as if the truth were to let itself be constrained by a pursuit by words. The illusion is that the stripping naked and the reiteration may turn the materiality of words into an event of truth. The cruel game, which at times is also conducted against ourselves, consists in hoping that inside or outside, and in any case at the end of a direction taken to its extreme, something happens, spirit materializes, corporeity transfigures itself, the hypothesis becomes fact, and the fact has the valence of an event. Here too, all this does not happen. Because neither the extreme of the spirit nor the extreme of the flesh can be more than themselves except than through an overturning that can happen only through the threshold. Words do not reveal by stripping something naked either; rather, they manifest while protecting, continuously bordering the threshold that they ultimately attest. This is the threshold of an unsayable that wishes to be said and of a continuously said that is not sayable.

Inside and outside are two modalities that are defined by the threshold. As such, they have an opposite and decisive meaning, and must be preserved in their opposition. Faithfulness to the impossibility of detaining both modes synthetically is faithfulness to the condition of finitude in which we dwell. The sharp positions that remain anchored to one side alone and are faithful to it appear to be more honest than the peaceful composition of those who consider themselves masters of both worlds. Such worlds are designed on the basis of the threshold, that is, on the basis of the place where they pass one into the other and one reverts into the other. Yet such a place does not exist. As we have said, [such a place] is not a safe and stable terrain in which one can dwell and from which one can begin. It is a place to be preserved in memory and sought through one's project. It is the most intimate place of all: the one that does not belong to us and yet is nevertheless ours. It is a tranquil and disquieting place because it is a place that protects but is not protected. It is a place crossed by negation but only because it does not quiet itself in an affirmation. It is the place that is one's own for everyone without ever belonging to such everyone.

The plane of spatiality, to which the concept of threshold is nevertheless in debt, never provides an adequate representation of it. Although the dwelling and the body are places from out of which one experiences phenomena of the threshold, they appear as marked by the prevalence of a "having" that does not seem compatible with the threshold. The home is the place of dwelling that one has and from out of which one designs an inside and an outside. The body is the physicality that one is and from out of which one designs an inside and an outside. In this way, the sliding of the threshold toward its being limit and barrier becomes inevitable because of the prevailing of a positive determination that does not do justice to the threshold. The distinction seems to be maintainable only at the level of the intentionality that one means to attribute to the line—which is; which one has. With a nonaccidental excess of emphasis, [one could say that] the threshold delimits (introduces, excludes) in a sacred manner, the border does so in a juridical manner, and the barrier does so in a factual manner.

TEMPORALITY OF THE THRESHOLD: PAST/FUTURE

Let us turn to time. Within it, one can ascribe to the present the same function that has been attributed to the dwelling or the body. The present is a time of threshold, a time of precariousness and crossing, the consistence of which lies in producing the overturning of past and future. Bergson's example of the sea wave, which he applies to creative evolution, can be helpful. The coast is certainly the place of the overturning, the point where the wave turns into surf. It is not its mechanical cause, though, which is more complex and multiple, and which implies not only the coastal form but also the force of the wave, the sea depths, and the wind, so that no wave is equal to another even if they all break and come back.

The present is the time of the overturning: an instant, and it is gone. Yet it is not a point, because the present remains, settles in the sand, anticipates the wave, and still resists its withdrawal, and at times even isolates itself from [the wave] until a new wave comes to gather the present that has no more ties with its past nor would have a future. One cannot halt the present. It is happening now; rather, it has happened. No, perhaps it is still impending. Yet when it has happened, the waters separate, like the waters of the Red Sea, and the past and the future configure themselves one on one side and the other on the other.

The present then is threshold. Let us recall what we have just said:[9] Like the threshold, the present discriminates between past and future, and like the threshold, the present is a fleeting zone, difficult to be determined even if capable of determining. There seems to be a perfect analogy with what has been said at the spatial level. Yet this is not so, and not only because, as one can see immediately, the opposition between past and future does not correspond perfectly to the one between inside and outside. The latter in fact designates a perfect alternative; within the former instead *tertium datur* [there is a third]. This *tertium* is the present. This has further consequences. In the case of the latter pair [the inside and outside], we have to situate ourselves within one of the horns of the alterative in order to experience a threshold; what follows is that, as we have underlined, a certain positivity that does not render justice to the threshold inheres to it. Conversely, here the threshold is said without occulting its dimension of negativity. For the present is exactly that which is not at my disposal. The past, albeit threatened, is mine; the future, albeit uncertain, is mine. The past is analogous to a patrimony, the possession of which must simply be exercised so as not to be terminated. The past is some sort of a storeroom of memories, which seems to be always at one's disposal. The future is analogous to a promise. We do not know whether it will be kept, yet it is a promise that is made to me alone in such a specific form. Perhaps it is only a dream, but it is my dream, the dream in which I mirror myself. In past and future we keep time at a distance; in this way, however, we also ensure that [time] remains at our disposal. How [can we] withhold the present, though, without its turning into a past that is threatened or its remaining as an uncertain future? My present is already yesterday; it is the threshold that I cross so as to take leave from myself. Or it is simply uncertain tomorrow, an anticipation of desire.

The present is not unless one lingers in it; this is also a feature of the threshold. [Such lingering occurs] not in the sense that one halts [the present], because the instant cannot be stopped except than in the dead fulguration of the beautiful; rather, [the lingering occurs] in the sense that one extends and narrates [the present], letting it pass and then returning to it, and accelerating it in order to then interrupt its movement. The present keeps escaping in the form of the flowing of historical time. At the second power, however, the present in which I linger and to which I return reflectively, as present and not as past or future, opens new regions and depths within itself. Its consistency, albeit fragile, strengthens itself; its fleeing, although it occurs repeatedly, delineates new contours. It no longer has the smoothness of a slippery surface but rather acquires roughness and thicknesses; in such a way, it is almost as if it adheres to the I.

When understood as threshold, the present acquires such an extension and depth that [such dimensions] can transform it into the turning point of times without thereby annihilating either the precariousness or the fleetingness that belong to it. If we did not root time and its passing precisely on this basis, then how could we understand that which is notorious to everyone—namely, that time passes and that, in the proper and authentic sense of temporality, it is never at one's disposal? Neither the past nor the future has such a true and full character because what is precarious is my relation with them, and not their self-consistency. Depending on how we consider it, their consistency is in fact either nothing (because everything has disappeared and nothing is yet) or intact (because that which has been cannot be revoked, and that which will be escapes me). In themselves and when they persist as such, the past and the future remain safe within their domain. A finite existence that is still in its fullness and is not subjected to the breakup of wear and tear, or a dream without realization—[these] maintain their own fragrance intact, like the unpicked rose. Dead or unborn events would preserve their precious form if the present, which is not and therefore must feed on past and future, were not to rummage among such material and take possession of it, thereby upsetting the already given form. Even the dead then are threatened, if the enemy wins; and dreams, too, can come back and move history, if the cause is just. If past and future truly and completely escaped becoming, then they would not be time but rather debris or stumps of facts. They, too, and not only my relation with them, are threatened. This happens because the present still lives in them, like still water waiting to get reunited with another wave. The present that is everywhere, the lingering that has escaped me and waits to be resumed, the interrupted yet not forgotten story, the narration left in the middle yet waiting to be finished are everywhere, and turn history into a time that remains yet unwritten.

The power of non-being that wishes to be is the secret of the present. This dimension, for which time does not belong to me, is the layer of the present that can be found in all past and future [time]. The past and the future no longer discriminate between themselves and the other. They intersect only in the manner of a reciprocal overturning. Past and future reciprocally interfere yet only in the form of an opposition. The past is submerged or explicit writing that cannot be rewritten except by another future that unfolds a forgotten meaning. The future is the inconsistent project of words taking material away from the past in order to exist. The present, which is not there, is the place where such an exchange of overturnings occurs and gets to be decided.

From here comes also the contiguity [of the present] with nihilism, which inclines precisely toward the present, or rather toward the instant. Yet [nihilism] does so only parasitically; it does so only in order to prolong itself, and therefore ultimately it is inexorably destructive (a subjectivity that does not get catapulted into an overturning leading it against and beyond itself inevitably ends up in a prolonged dying). Conversely, non-being longing for being is an instituting force; it reverts and guides, takes possession and gives consistency, scans times and ensures strength. In its own way, the finitude of the present too is a barrier; [it is] the origin of a negative that wishes to be and therefore gives rise to extension and depth. Non-being makes being possible. The present that does not settle in the form of presence demonstrates that for being there is no guarantee of necessary permanence and stability. Through the present, the moment and threat of negativity insinuate themselves in being; it is the present that may fail and get lost, the present that must register a lack and finally an absence instead of certifying stable presence. Yet there also dawns the possibility of giving an albeit precarious and nongranted place and time to that which longs for being.

The issue that presents itself from out of what has been said so far is that of understanding that, if the present is an eminent figure of the threshold, then time is not simply an arrow inexorably cast forward, and the past and the future can even turn one into the other. Conversely, our entire symbology of time considers [time] as a line of continuity, in which from the past one arrives at the future passing through a present the consistency of which is variously understood. The difficulty that arises from this has to do with recognizing its own nonaccidental specificity to each temporal modality. If linearity were the scheme through which one would represent the scale of time, then the present would be the accidental moment on the basis of which one could separate past and future. At the same time, by moving the ruler of time further backward or forward, that which was future would turn into past or that which was past would turn into future. In this way, though, the distinction among the three modalities would be entirely accidental; in the end, it could only find its rooting as a function of the observation point, as Augustine, in a certain manner, shows. The further consequence of this is not only that time would not have its own consistency, but also that features such as being-past or being-future would subsist only as modalities of the subject without their own factual reality. In such a case, only a sort of preestablished harmony of a Leibnizian kind would be capable of justifying everyone's simultaneous belonging to the same present at a given moment, and as a consequence the very discrimination between future and past. Nor is it possible to return to an objectivistic conception of time, in which flowing as such produces the passed feature of the past and indelibly assigns to it an irrevocability that would constitute its characteristic notation.

Through the notion of the threshold we have tried to retrieve first of all the sense of the present, its specificity as a zone that escapes me but that (the zone, and not the subject!) provokes effects and consequences. Reciprocally, through the concept of the present we have found a figure of the threshold that is more adequate than all spatial figurations. This is a precious indication confirming the ontological primacy of temporality.

Let us recapitulate this once again because the thematic cluster is important. On the one side, there exists only the present because it is its experience as threshold that attributes to time its feature of fleetingness and implicates the subject in time by situating it within [time]. To say time is to say the present, which is other than, if not even the opposite of, presence. On another side, time is only the past or the future, because the present is nothing else than a divide between the one and the other; it is the instant in which the subject leaves or anticipates. On still another side, as Augustine already knows, the subject is time, yet without full coincidence [of the two] because the temporality of the subject is not the projection of an imaginative scheme modeling a formless nature. Instead, [such temporality] is the essential extraneousness of the subject to itself; it is that which, by stretching the punctuality of the I, entrusts [the I] to that which does not belong to it and even escapes it, to that which fills [the I] up and even consumes it—that is, to time.

The present is pure temporality, and thus time that I do not have. From this perspective, since it is that which withdraws itself, [the present] also appears as non-being (and we have called it such): it is not presence, it is not eternity, and it is not objectivity. Not for this reason is it subject to manipulation; thus, it does not reduce itself to a subject's perspective on temporality (in fact, it is time that I do not have, whereas I solidly possess the past and the future). It is nevertheless the time that is to me [*tempo che è a me*], that invades me, and that is given to me. The present is today: for everyone, it is a "to me" ([but] not thereby simultaneously a "to us" for all of us). The struggle of the subject with the present is that through which one arrives at the constitution of time in the form of history, something through which this non-being that I do not have but that is only "to me" comes nevertheless to being. The success of such a constitutive expansion of the present into an unfolded experience (which is the success first of a fragment of life and then of the whole life) does not automatically guarantee that my own present becomes the present of collective history. This sharing is more complex and indeed not guaranteed. Truly happy can be called the life of the one who succeeds in proposing the "to me" as a "to us" that can be shared not through the illusory projection of the self onto the collectivity but rather as the donation of a time that can be shared. Through this path, within the frame of a "to us" in which we take part, not shelling the today dispersively but rather appropriating it as a today also "for me" becomes easier for everyone.

AFTER MODERNITY: THE REDISCOVERY OF THE PRESENT

One can speak of the present in this manner only after modernity (an after-modernity that does not coincide at all with postmodernity). Modernity is the time of the new for the sake of the new. Thus it is the time of the interruption [*cesura*] that the new, insofar as nonexistent, introduces into history producing its transformation. In this sense modernity is the time of the future, because the future incessantly drags in the direction of replacing the new with a more recent unheard-of novelty. In modernity, the pressure of non-being dominates everywhere and accelerates history toward death. Heidegger's primacy of the future is the perfect transcription of this; with [such a primacy] we are also presented with the nihilistic seal of the dominating conception of history. Progress is much more an escaping from life than an augmenting it.

At the origin, when the modern made its appearance without yet being modernity, the theme of the new for the sake of the new was meant to account for time and history. It was meant to be and was, in fact, a rebellion against the immutability of nature. It was even meant to account for the birthing of the new, and for its unpredictability. In all this there was much generosity, a trusting and open spiritual attitude—the desire to let pure *Erlebnis*, pure impression, surprise existence. *Erlebnisse* and impressions are infinite though. They are minuscule, almost imperceptible, but infinite. Their uncontrolled multiplication has escaped the hands of the generous initiator of the modern.

The infinite opportunities of the future have given rise to a future that is rather made of unknown instances. Anxiety and then *Angst* over existence have taken the place of the secure generosity of spirit. The future has not constituted a history; conversely, it has inexorably eroded it.

The modern and modernity are at the end, however, and a different, albeit uncertain time presents itself. [Such a time] could build something precisely by starting over from the present, from the act through which the present makes that which remains emerge from out of itself. [Such a time] does so by recovering the legacy of a past or the project of a future.

Such could be the new historical task of Europe, which is not a land of the sunset, but rather a cradle, a place of birth. With its wide-open spaces and [as land] where luxurious nature prevails over the choices of history, America can still and perhaps always hope to give new welcome to the latest and further novelty. [Being] so narrow and crowded, and where it is history that makes nature beautiful, Europe must expand its place of consistency. Its future is in the present, in the valorization and trepid articulation of diversities that are rooted in history, and that must be subtracted from oblivion without thereby making the conflict intolerable in order to accomplish [such a result]. To give voice to differences, to protect and reconcile them means to protect the narrow [available] space and offer a new time. It means to bet on the present, not so much to stop it but so as to constitute it (as institutions constitute; politics, as Aristotle reminds us, is architectonic knowledge). Only the present donates (*stiftet*) time as the human time of finitude. Temporality alone makes comprehensible (and reconcilable) the paradoxes that make up Europe and its wedding the existing identity (its history) and the existing differences (the history of its conflicts). Only when they are scanned through time and are not purely unfolded in a system is it possible to account for the one and for the others, and understand that the *proprium* of European culture is in its not having a *proprium*, as many have remarked. This is not a form of weakness and indifference; rather, it is an attitude that, in the long history of Western culture, has been appropriated not only in order to hold on to its own viewpoint but also, without abandoning it, in order to understand the reasons of the other, to look for the other so as to recognize the other and be recognized, always there where one is not. Only when they are scanned in time as present time do these paradoxes obtain a nonabstract conciliation, for they are obliged to a confrontation with the urgencies of today and with the need of taking a stand. Among the many perhaps possible ones, Europe becomes that interpretation of the West that finds an unthought-of timeliness in the today because it is not simply an ideal hypothesis but also a historical, albeit marginal sketch. This, of course, offers no assurance of success; it only deepens responsibility, that is, the ability to give an answer to those opportunities that history presents, today and not forever.

We need a new myth: not in order to escape but rather to resist. After the kidnapping that violated her, Europa could come back, take the veil of deceit away from time, and give [time] back to us as present, as the undefeatable place where to stand. Zeus's gifts to the kidnapped maid, who was taken away from her world of games, were instruments that never fail their goal, surveillance over the Cretan coast, and the individuation, and catch of the target of the hunt. Removed from the ingenuous flowing of time, Europa seems to have obtained the gift of ensuring infallibly for herself that to which she aims.

COMPLEXITY AND CONSISTENCY OF THE PRESENT

The present cannot be deferred. From the past, one could simply take distance and the deferral would then take up the form of disengagement. With respect to the future, one could always take some more time, and the deferral would then assume the form of dissatisfaction. The present, however, cannot be deferred, that is, put off to a new role. Perhaps it can be prolonged through lingering, which expands [it] and gains new time. Lingering remains within the present and prolongs it. Conversely, deferrals feed on indifference. Because all things are the same, times too are the same, and everything slides away and is deferred. The present is instead here; it impends, and it demands a subject that handles it. The time centered on the present is not cosmological but rather human time, which presupposes a subject (not a master) of time. This subject that ventures through time is no longer the solipsistic subject. Having risked itself in existence, having entrusted itself to temporality, the subject has encountered the extension of the self. Yet this extension that, as the figure of the threshold has shown to us, always occurs within an alternative between oppositions (outside/inside, past/future) unfolds but not like an inertial expansion. It knows oppositions, fatigue, and rough rubbing. In principle (without wishing to account properly for this here), the subject from now on inevitably finds itself confronted with its own and the other's alterity; thus it is open to a possible intersubjectivity first of all because it is indeed implicated in a tangle of intersubjectivity.

The present demands a complex subjectivity, something that is not reduced to a nonextended point, a mere principle or an extraterritorial foundation. The present itself does not appear as a nonextended punctuality (or a punctuality that can be extended only through a linear and repetitive prolongation), but rather as a complex and intertwined tangle of times. The present is a risky and ambiguous opening, which is simultaneously conditioned by the past, which it archives or for which it longs, and open to the future, which it fights or for which it fights. Every present, that is, everyone's present is like this. Yet no present is present only to itself. The present is exposed to the present of others and interrelated, even interwoven with those. No present is therefore simple or can simply be reduced to a single subject. For in it subjects press up and interact in the same manner that times press up. Human beings who are abandoned on a desert island, human beings who are absolutely alone, without relations and unable to build a history, are without time because such human beings are without present. They do not lack the past because no one can take the past away from them. Nor do they lack the future because dreams (if they had a present) would not be precluded, nor would projects be forbidden to them. Yet the present is removed from [such isolated individuals] because for them the succession of instants does not succeed in building a field, in becoming threshold. There remains only an individuality folded upon itself, and the succession of an astral chronology. For them, the present gets to be progressively reduced because they construe it in a residual manner by taking desperately from the past and creating a glimmer of future from out of that (and not from out of their own present). Were [such individuals] to give up all ties with their history and all expectations for a change (the salvation that comes to them from others through the immense sea), they would no longer know time, nor would they have any present available, not even the residual present of which we spoke. Is this still a present, when it is so contracted and subaltern? Has it not lost all the vigor and force that make it precisely the present, that is, a lively existence projecting itself and seeking prolongation?

To start again from the subject, to start again from time, to start again from the present. To follow the folds, the places of the failure and the sudden exultation of a success; to reweave a knot, and narrate it. To start again, after modernity. It is clear that the simple philosophical enunciation of the subject as the place of consistency from which to start, or the discovery of the present as the time that allows a complex unfolding of the subject, or even the identification of the threshold as the turning point capable not only of producing overturnings, but also of giving raise to its proper zone and dimension, are by themselves not enough to guarantee such a beginning again. Philosophy does not produce anything, yet it can prevent much. It does so when it thinks not only about the mode of time in which it lives, but also against and therefore beyond such a mode. Only by doing this does [philosophy] make a beginning possible again because it institutes the conditions of possibility that render the new beginning precisely possible. It seems to me that the present, the subject corresponding to it, and the threshold that is its dimension may be such conditions of possibility. They do not produce but announce a new beginning.

I repeat: The present is the unique threshold on which every existence is built. The present is the condition capable of accounting, by casting light on it, for the attempt at extension that belongs to each existing. To use a metaphor that comes from nature, and that for this reason must pay some additional dissymmetry in comparison to other metaphors: The present is like the Western archipelagos of the Dutch islands that, because of the orographic nature of the ground, change their configuration daily, yield portions of beach to wind and sea, but in other points also gain new coasts precisely thanks to wind and currents. Will to resistance and ability for new adaptations are implicated in the present that configures and reconfigures itself daily, and they constitute its strength.

CHAPTER THREE

Ethics of the Present

EDUCATION AS ACCUMULATION OF PRESENT

A not irrelevant confirmation of this attempt at defining the present may also come from that operation of formation within temporality that is education. Unlike training, which is exclusively functional to the results (and hence to the future), education is an attempt at gaining such a consistent ground of present that [the present] turns into the sufficient base even for the simply unexpected. One should keep well in mind that results do not amount to success, which is a term with ethical and ontological connotations. Success evokes a fullness that gives itself the form of a present; results are simply congruity with an objective. There can be success without results, and results without success. Success is a form that may even not result in recognition. Results may depend on benevolence of fate or cunning of one's moves, yet they do not receive from this any guarantee of instituting a surplus of reality. In this sense, an education geared at forming in view of something is always doomed to failure, either because what presents itself will be different from what was expected or because simple repetition does not guarantee automatism of acquisition at the moral level.

Education tries to accumulate experiences. This does not amount to storing a past from which to draw (as such, [the past] too appears as nonutilizable: whatever one may say, history has never taught one anything). Rather, it means nourishing oneself with the present, precisely like a battery that charges itself with energy. Education means accumulating such a powerful and rich present that one becomes capable of acquiring ample strata of past and future—the others' past history, which is rendered mine and timely, and the future, which remains unknown and unexpected and yet can become something that, in [my] projects, already belongs to me and even directs me. Successful education means ensuring a ground where forms of continuity can be generated out of the awareness (experienced to the point of pain) that an interruption has occurred, the tradition is interrupted, and discontinuity is the normal condition. Aware of this, education nevertheless works for a possible and shared continuity.

Let us be sincere: *Bildungsromane* are simultaneously pedantic and boring. *Wilhelm Meister's Apprenticeship* still suffers too much from the Enlightenment framework to be able radically to address the theme of formation. The *memento vivere* [remember that you have to live] that resonates in it is thus in singular contrast with the artfulness of the path to which, precisely in order to enter life, one should submit as if to an initiation. In this sense, Alessandro Manzoni's *The Betrothed* is much deeper; it makes all the vicissitudes of the novel originate from a present that has been impeded. The violence that disrupts the dream of a present (that is, of an engagement extending into marriage) leads Renzo to Milan and transforms him into a participant in the vicissitudes of his time.[1] These [vicissitudes] are no longer—abstractly—trials; rather, they are obstacles. The difference is substantial, like the difference between tests and life. The story, humble and simple, intertwines with the big history. It does not flow into it, as happens too optimistically in Goethe. Rather, it crosses it, and in the end (the happy end, which is too trivial to please) it returns to the present from which it had departed: Renzo builds a family after he has seen the world. Whether or not one likes it, all this is deeper than what happens in Goethe. It is deeper precisely from a pedagogical standpoint. It has fewer illusions, it is less universal; yet, it knows that what makes human beings greater is not their many past vicissitudes but rather their ability to assume them as an element for widening and solidifying their own present. Without being pedagogical, Manzoni's novel intuits that the present cannot be overturned but simply widened, as is shown by the modesty of his conclusions, which appear annoying for too much leftist ideology (which then, outside totalitarianism, pays the price of its own irrelevance in everyday life).

That education is successful can be said only afterwards, only at the end, once the path has been completed. For nothing is ever truly secured. Where the threshold that has been crossed may lead can be known only after the decisive step has been irremediably taken.

THE PRESENT OF THE INSTITUTION

From this perspective, one can understand also the danger of and need for institutions, which are forms of dilation of sense, present, and intersubjectivity. Institutions (family, society, and so on) constitute an opening in order to secure intersubjective relations to the present by articulating them in a horizon of meaning. It would not be correct to assert, as it is normally done in a polemical, anti-traditionalist perspective, that institutions are a thing of the past. This is, of course, true for those institutions that have lost all propulsive force; yet, this does not affect either the institution in itself as such, or even less the fundamental need for instituting.

All institutions are historically dated. This means that they have had a beginning and through their own institutionalization they have tried to secure a present extension for such a beginning. In this sense, lively institutions are always places where time circulates and the fatiguing effort is at work in order to secure the present the double register of legacy and projectuality. The risk of institutions is, of course, that the inertia of the self-reproducing existent may prevail over existence that fills itself up. That is, [the risk is] that the precarious balance between the institution that entrusts itself [to us] as task and the individual who receives, vivifies, and transforms [such an institution] may get stuck (beside occurring with time, this may, however, be a defect already there ever since the beginning). Where the individual and the institution do not exercise each its own rights, where they no longer engage in that close struggle [*corpo a corpo*] that produces the normal dialectic of history, both will turn out wounded—the institution in the form of its becoming sterile, the individual through resentment. Yet outside this struggle, which is the measure of history and the condition of change, the opposition between individuality and institution is on the one side superficial, and on the other completely mistaken. It belongs in fact to the interests of the individuals, it is even their ambition to prolong themselves in the form of an institution. What at the surface or superficially appears as the prevailing of the general over the particular is rather the imposition, over a purely existing individuality, of an individuality that has turned itself into a historical institution.

Beside the factual historical outcome, the problem becomes irresolvable without an articulated philosophy of time. Where the nobility of the past prevails, one cannot escape the venerable primacy of the institution except in the form of a compromise or an exception (the reactionaries always provide also for the state of exception; to justify the historical changes, from time to time they allow for a *tabula rasa* on which to reconstruct and start again to preserve; reactionaries and revolutionaries—both detractors of the present—not infrequently even exchange roles, as Carl Schmitt and his reception teach). Where there is only expectation of the future, the institution is suppressed in principle yet provisionally admitted in view of its final and universal suppression (hence it no longer leaves room for criticism and beneficial dialectics because it is absurd to conflict with what does not exist except as a ghost, even though through such a simply ghostly existence it functions in an concretely oppressive manner).

Only the primacy of the present, as time that institutes within precariousness and gradually augments itself through a dialectics with existed past and expected future, enables the ordering of the relation between individual and institution, according to an ontologically diversified yet historically interwoven scheme. The individual's rights, albeit less extensive, antecede and outlive the institution in the sense that they come before and continue after it. Therefore, in the case of a radical and incurable collision (one that could jeopardize the sense of the institution and therefore the sense of that which orders sense), the individual's minimal rights cannot not prevail. The institution cannot claim a value in itself (which would transform it from an extension of the present into a form of eternalization of the present, that is, into a presence). Conversely, the in-itself of the individual does not surge to being a value unless it extends the threshold of its own present into a crossing with intersubjectivity within a mobile horizon of meaning, which is also structured in the scheme of an institution.

This dialectics is simply the exemplification of a more general model. The individual's minimal rights hold less value, but they count more; the maximum rights of the institution hold more value, but they count less. With "value" we mean the extension, the quality, the intentional content. Elsewhere this will be called "sense." Value and/or sense are constituted within relationality and history, within the extension and endurance of the claim [they make]. Yet they (which are *maximi*) are not originary. At the origin is the little, the stubborn, the *minimum*—the self-reproducing individuality. It counts more because one cannot go below it. One does not bring individuality any advantage by naming it "value," or considering it in terms of sacredness. One can rightly consider it from above, with no need for rhetorical overevaluation; nevertheless, it remains as a fixed point, a minimum of no-return. It forces one to humility, to come to terms with the stubborn existence of pure existence. When one suppresses [individuality], one suppresses sense [*senso*]; whereas when one takes sense [*senso*] away, there still remains the individual, the punctuality, or that which elsewhere will be called meaning [*significato*]. Hence comes the correctness of the saying *primum vivere deinde philosophari* [first living then philosophizing], even if it is philosophizing that makes life worth living.

For this reason, the greater dignity of the institution must always step back (a step after the beginning, a step before the end) in front of the individual, whose minimal rights are inalienable. The most fundamental rights are also the most banal: to live, eat, be dressed, have children—the expected rights of everyone. Nevertheless, without such banality, which is life, no deeper value can be built.

FOR AN ETHICS OF THE PRESENT

One should delineate a map of the virtues of the present.

The treatises on passions, which have dominated the modern age, are an attempt, which is also political, at constructing a virtue of the present. They neither seek the eternal nor do they define an absolute order of morality; rather, they describe the human condition in its difficult balance among different impulses contending over human nature. Apparently atemporal, like an anatomical table of spirit, they are nevertheless well aware of immanent temporality. Despite all, remaining rather far from the Stoics, who taught to repress rather than govern the passions, they help to handle [the passions], from Descartes to Spinoza. In Spinoza, then, the free person avails oneself of a wisdom that is *meditatio vitae* [meditation on life] and *non mortis* [not on death]; one is cheerful and joyful, and pursues the increment of such joyfulness as the passage from a less to a greater perfection. Everything unfolds through a *conatus*, which is loaded with time because it produces an increase in being, as Aristotle's search for happiness already does. The appetite leading human beings and structuring their passions is geared toward the increment of being; that is, it is present time seeking its own extension and consolidation.

To turn backward toward such austere treatises is of course like referring to the remnants of a humanity that has disappeared: the human being described here is too evidently an ideal-universal human being for us to be able still to identify with it. Such treatises remain austere and venerable, but they speak to our heart with difficulty (and for this reason perhaps they are usually assimilated to the Stoic ideal, which is precisely a past and nonrepeatable, even an atemporal ideal).

All this has probably to do with the kind of present that such treatises enable to save, and with the purely defensive attitude that characterizes them. The strategy at work in order to dominate or master or channel passions is certainly first of all an attempt at excluding the interference of the future or the past, as is shown by the two foci of hope and fear, against which one equips oneself so as to be protected against that which escapes domination by reason (the inadequacy of knowledge, the phantasm, or the corporeal).[2] The domination of passions turns out to be well focused on the present, even obsessively turned toward it. Its outcome, however, has the form of shrinking the present. The dominated passion is an effort that is carried out entirely within the subject; it produces a concentration on the present, whose price is, however, the decantation and rarefaction of the elements that compose [the present]. The present of the dominated passion is purified of passion. What remains is a corrected, simplified, more rigid, and frailer present.

In a sense, this present is a time purified of time. It is a time that has turned elsewhere, toward the care for the purity of existing that consequently loses its own consistency. One could say that such a present has the purity and transparency of a crystal, because it is the attempt at protecting the present against any contamination. Not unlike this is also the political usage of the purification from passions, as the Jacobin Terror exemplarily manifests. What results is a pure present, even an absolute present, which not only is not between times, but also excludes all relations among times. In this way, the present ceases being that which discriminates between times and becomes a residual time, because the present as debris, when it is absolutized, risks turning into a monster. Likewise, the virtue that is obtained through the impoverishment of passions is a residual virtue; it is what remains through an operation of self-domination (and this too is no less unnatural and therefore monstrous). In Aristotle, virtue was still the mean [*medietà*], that is, that which measures excesses and judges of them as such; here instead virtue folds upon itself and produces the expulsion of passions.

The crystalline purity of the human being without passions is the instantaneous conversion of finitude into infinity. It is the present gained not through extension but rather through folding. It is the present that is all surface without depth. It is the present that, by not being between times, has no roughness and avoids the negativity of evil because it simply expels it. The treatises on passions describe human beings who are all present, yet present to themselves. They gain the present through the subtraction of any other time.

In truth, such is not always the case, as the reference to Spinoza's treatise has already hinted. In Spinoza, the present appears as tranquil and cheerful, slow and expansive strength. Nevertheless, even such a modality, which is more trepid and human, is no longer [fit] for us because for us the present is precarious and uncertain, sudden, like Benjamin's dialectical instant, and threatened. Such a latter present, which is ours, appears even fuller with time to the point of explosion, of becoming interruption that saves, of assuming apocalyptically revealing features. [To say it] again with Benjamin: It is the "emergency brake" of revolution, that which does not speed up the acceleration any further, that which rather halts it, and brings it back to the day of judgment.

The present, this present that is tenacious and humble, threatened, and precarious, yet stubborn and capable of self-perpetuation, will halt modernity and will be able to save it. The collapsed Twin Towers will not need to be rebuilt bigger, nor will they have to give rise to a forward rush of new punishment and absolute justice. Instead, they will have to remain like a hole that cannot be filled, a snapshot of the present that reveals to us not only the present in its potential madness but also in its solid roots.

These are the roots to which [one must] return and build, not in order to go beyond or elsewhere or in the direction of a constantly renewed new, but rather in order to help us stand where we stand, to find ourselves where we are, and to consolidate the brief space that is existence. This space may no longer be sacrificed to history with impunity; history may no longer tribally take over existence, revenging itself over it, and receiving from its halts new material for its very own accelerations. The present will halt modernity. A gesture is sufficient; and the incessant movement of time escaping itself can be saved.

The need for change may arise from the perception that salvation has been lost: one needs then to convert, turn around, and retrace one's steps. Or [such a need] may arise from the satiety due to the salvation that has been found: it is so great, and it measures me so much that I cannot stand it. The present can halt time. Yet [such a halt] need not be an apocalyptic conflagration or a revolution, nor should one think that everything is redeemed or lost. [One needs to] halt time so as to give something the time for self-consolidation and being, and for being able to continue to be also in the future. Perhaps one does not even need much; perhaps what is needed is just some patience. We do not know this in advance though, and experience no longer teaches it to us either. Neither does knowledge of the past teach it to us, nor does the secure path of future progress indicate it to us. It is rather taught by a babbling and recently learnt language; it is the language that composes words in the present and lets them resound, so that they may take up form and consistency, and so that other words may be said with respect to them.

Such time certainly knows itself as an in-between, not as an absolute. Being removed from the linear cyclicity of succession, it has not thereby become unrelated (in this sense, it is different from the naive systematicity of the treatises on passions). It knows itself as intermediate by nature; yet, it knows such intermediacy not as the result but rather as the source of preexisting magnitudes. The difficulty of a time that comes after modernity is that it must join the present as in-between with the posing constitution of the present itself. It must reconcile the determinateness and limitedness of the thing with its own founding nature. It must measure finitude and infinity, the conditionality of the present as the time of finitude, and the absoluteness of the present as (more) originary temporality.

New virtues and new treatises on passions are required for all this. I do not dare venture into such a direction, which demands a long and thoughtful moral experience. Nevertheless, I think that I can say, as was already the case for Spinoza's *conatus*, that the virtues that we need must contain a noble magnificence, a sense of and will to augmentation not as a form of an ultimately desperate *Streben* [striving], but rather as a generous and communicative positivity, a contagious and spreading abundance.

The virtues we wish to seek move from the present and widen its consistency. They should perhaps be sought in the midst of *hilaritas, humilitas*, generosity (*Großzügigkeit*), and Bloch's daydreaming.

Spinoza's *hilaritas* goes in this direction because it is a "spontaneous force within us,"[3] and one simply needs to know how to direct it. It is a diffusive joy that longs for communication. There is much that is evangelical in this—much of that cheerful and surprising spirit that amplifies the announcement. [Paul's] epistle to the Thessalonians is a singular example of such an impatient cheerfulness. It is certainly pervaded by an enormous expectation for the *parousia* [Second Coming]; yet, it does not thereby pass over in silence the human need to see one's own brothers and sisters again, to mirror oneself in them, and thus to have one's own joy and glory already now. That is why, all expectations for an imminent end notwithstanding, Paul does not fail to exhort his brothers and sisters to augment the present: "We exhort you, brethren, to do so more and more, to aspire to live quietly, to mind your own affairs, and to work with your hands, as we charged you."[4]

Hilaritas is not self-sufficiency in happiness, that is, something that nourishes itself on itself and draws from itself. Were it so, it would end soon, and in any case it would not augment. *Hilaritas* is a joy that I receive and therefore transfer in the form of a gift. For this reason, the gift cannot be simply transferred beyond, passed from one's hand into another's; otherwise, it would become an exchange good. The gift, on the other hand, in its characteristic of being a surprise—a characteristic so essential to the gift that the gift cannot be asked but only waited for—is a nonexchangeable, absolute act; it is a surprising generosity of being, which in its innermost essence is joy, *hilaritas*. As Jean-Luc Nancy remarks:

> The gift is . . . that whose "present" and presentation are not lost in a realized presence. The gift is what comes-up to the presence of its "present." It also keeps itself, in this coming-up and surprise of the gift, as gift, as giving of the gift. In this it is an offering, or withdrawal, of the gift in the gift itself: the withdrawal of its being-present and the keeping of its surprise. It is not a question here of the economy of the gift, where the gift comes back to itself as the benefit and mastery of the giver. On the contrary, it is a question of what makes the gift as such: an offering that may not be returned to anyone, since it remains in itself the free offering that it is (that is why, for example, one never gives what one has received to a third party, lest one annul the gift as gift). One must keep the singular present in which the gift as such is kept, that is, offered: it is presented, made freely available, but is freely held back at the edge [*bord*] of the receiver's free acceptance. The offering is the inestimable price of the gift.[5]

The attitude accompanying gift-giving can be transmitted, however, and it is diffusive and contagious: it is the generosity of cheerfulness.

Evangelical and then Franciscan *hilaritas* nevertheless stands well rooted in the earth; it is *humilis* [humble], conscious of the lowness of creatures, as Luther underlines when commenting on the *Magnificat*, which he translated into German (1521). Humility and cheerfulness are strictly connected because the very awareness of the lowness of creatures emphasizes the surprise in the gaze with which God addresses such creatures and looks upon them, thereby raising what is low and abating what is high.

Before being a moral virtue, humility is a fact. It is being small, close to the earth, passionate for what is little. The strength of such humility, of this kind of humility, is upsetting because it institutes a new logic. It is like looking at the world from below rather than from on high. One can see things that the one who stands too high cannot see. Moreover, it completely changes one's perspective. The humility of God's servant thus becomes the sovereignty of the woman who is the mother of God, and the wind coming from such an overturning in the criteria of judgment overthrows the powerful from their thrones and raises the humble; it sends the rich away with empty hands and sates the hungry.

Marian iconography is the representation of a woman, small and nourished with the materiality of the earth, who proudly presents a child who has been promised to her as the son of God. Everything is *humilis*, but her gaze is not modestly cast toward the ground. She looks at the child to protect and withhold him, and to comfort herself; yet, she looks, beyond the canvas, toward the interlocutors and sustains their gaze, because she presents them with a different order of values.

In Christian representation, God has to bend to the suffering of humans so as to speak to them. God himself has to become humble so that his generous lordship could become attainable. God, too, has looked at heaven from below, even from that abysmal lowness that is the Cross, from which the Father appears faraway and incomprehensible. God had first of all to convert to the Cross so that the design for his creation could be complete.

Humility, which we learn with such difficulty, is pride in and passion for simple normality—to be mothers, fathers, and children; to rejoice in what is given to us, to exhibit it with the certitude that it belongs to us not by right but as a gift and that it is therefore above all rights. Thus it equalizes even the unequal and lowers that which presumes to have greatness in itself.

Valorizing that which is humble and hidden, simple and without pretense (and therefore also hilarious, full of joy) requires generosity and openheartedness. It requires a kind of nobility that is not aristocratic but rather a mark of simplicity and purity. Generosity presupposes abundance:

abundance in measuring—the greatness of that which does not reduce what is under consideration to small categories, to reductionistic grids—and abundance in giving, both in terms of attributing recognitions and rewards and in terms of assigning roles and spaces, and delineating horizons. For this reason, the German language, in its linguistic characteristic of being materially concrete, expresses the concept of generosity with a truly successful word, *Großzügigkeit*—that is, that which has greatness as its constitutive feature, a greatness that measures only in terms of largeness, according to the norm of abundance. Generosity is the greatness and overabundance of being, the measure of which is accomplished today, and neither tomorrow nor in the future or when the moment comes. The moment is today. German train stations are furnished with improbable clocks that, when they reach the minute, stop for an instant before starting again counting the seconds—as if there were a donated time between one minute and the other, between an hour and the next. This instant of pause marks the minute and the hour; it marks fulfillment and birth. Every time, I intend to look at such clocks carefully because they mark the time when a circle turns itself into another, when the end goes back to being the beginning. They mark a time of salvation.

Generosity is magnanimity, that is, a greatness of soul that arouses and expects further greatness. It is the ability to measure according to a great and open measure; therefore, it is always demanding, never petty, obtuse, and fiscal. [Generosity] pulsates with the greatness of the present and therefore it expands it toward wide horizons.

Daydreaming, as Bloch speaks of it, is related to [generosity], because [daydreaming] attests an overflowing of the present toward the not-yet-occurred. Of course, as Bloch himself acknowledges, there are Hollywood dreams, which have been created precisely to delude and distract from the misery and injustice of this world; and then there is daydreaming, that is, the not-yet-realized yet realizable utopias, the hopes to which the heart, through its anticipations, already gives the contours of reality while thereby helping us to pursue them.

The distinction between diurnal and nocturnal dreams has to do with a fundamental feature of daydreaming, namely, that [daydreaming] does not produce the obliteration of the ego; rather, it holds on to it. Whereas opium, as Bloch remarks, produces a dulling of the I, hashish alters the I's functions yet without weakening its consciousness. Likewise for daydreaming, which aims at fulfillment, and has world improvement as its goal; [daydreaming] does not end up with fantasies because it does not relinquish its reference to the I from which it originates.

Dreams are desires, as Freud also knows. Yet, at least to my mind, desires are not always correct; they should not always be pursued—hence the double, ambiguous character of dreams. It would be naive to claim that one should dream, and it would be unjust to say that one should not dream. One should in fact daydream. One should connect desires with needs, with something that is missing and should instead be there. Not only desires for power, greatness, and success—that is, solipsistic, self-centered, adolescent desires—but dreams that are desires for that which is missing, for the place where we have never been but that is ours: transparent dreams, in which I can reflect myself and which I can share. Dreams are also collective desires, because they are needs of which others partake.

Children's dreams are grandiose, yet not excessive; they dream that what they have—that is, the affection surrounding them, the joy they have experienced, or the home they live in—may be the way it should be. This is an enormous dream, yet not an excessive one. They do not dream something else than what they have; simply, they dream its absolute realization.

One may not dream less than the absolute, yet one should do so while remaining within the finite measure that is given to us (this sounds like a contradiction, but it is not). Since adults are ultimately realists and a bit disillusioned, they complicate their dreams. As a matter of fact, they do so out of realism; they no longer dream the absolute but rather the means that they deem necessary (and attainable) to reach the absolute.

One should demand more, and dream the absolute. One should fall asleep and dream God. Perhaps one would then wake up and discover that the dream has come to fulfillment. As Benjamin writes, "the fairy in whose presence we are granted a wish is there for each of us. But few of us know how to remember the wish we have made; and so, few of us recognize its fulfillment later in our lives."[6] Let us hope that we are among those who will have realized this.

Daydreaming indeed anticipates the future, yet in the sense that it compresses it in the form of a present that unfolds within the known span of origin and realization. It gathers into the space of the present the not-yet-lived [understood] not as an anticipation of a simply possible but rather as the explication of a real that is not yet entirely known. For this reason, [daydreaming] is also marked by a problematic overabundance, that is, an overabundance awaiting its realization. For this reason, it rightly takes up an ethical value (the goal of ethics is, following Aristotle, the search for happiness), and inscribes itself among the virtues of the present in their upper margin, that is, in the margin closer to the future.

BEYOND VIRTUES: THE REASONS OF TENDERNESS

To understand the reasons of tenderness is not difficult. The virtues we need are in fact "gentle" [*dolci*] but not weak. Moreover, they are not simply moral virtues in the sense of concerning "how to act" or "how to do."

Rather, they define our relation with what is, and choose the being [*essere*] toward which to orient ourselves; therefore, they have a strong ontological value. Tenderness is the ontological sentiment par excellence.

Like all feelings, tenderness has to do with the intertwining of being and appearing. More intrinsically than all other feelings, however, it considers appearing not as the mask but rather as the relief of being, and it concentrates on appearing as the manifestation of a form of being. From this perspective, as has already been remarked, [tenderness] can be considered as an ontological feeling because not simply does it connect with being in a general sense, it also shows how its historical self-giving, that is, its appearing, is essential to being. In the way in which tenderness gathers it, appearing is not the opposite of being, according to the canon attributed to Parmenides; rather, it is the manifestation of being. What emerges in this is a conception of appearing not as an overabundant (and ultimately inessential) expression of being, but rather as its determination and destiny. The fact that being is entrusted to appearing tells an essential feature of being; it emphasizes the essential historicity that runs through its nature. The mode of appearing is not indifferent to the being of being [*l'essere dell'essere*]; rather, it discloses its innermost essence. For this reason, one regards the being [*l'essere*] that appears with trembling spiritual suspense, the expression of which is tenderness.

Tenderness gathers the fleeting frailty of the existent, which it sees as threatened and transient; yet, it does not thereby situate it under the sign of death. In fact, [tenderness] does not gather the warnings of decay but rather the surprising freshness of life. It does so with the penetrating gaze of one who walks through the complex and multiple layers of reality, and who therefore understands appearing on the background of being, and being in the relief of its manifestation.

Tenderness gathers an intertwining that as a matter of fact can always be untangled but that, at the level of that which is valuable and deserves being, should never be lacerated. Tenderness concerns the present. It cannot concern either the past or the future except than insofar as they can also be present. Tenderness concerns the frail stability of the present. It reads time between times. It reads the in-between-times of time, that is, the present. And it reads it between times: between itself and the future (will this present resist the insults of time?) and between itself and the past (will this present have enough strength not to lose its ancient roots?).

In correspondence to such an intertwining, tenderness highlights an attitude that comes from the ability to look at things as if through a double filter. The analysis of experience makes this very evident. The child is an eminent object of tenderness. Tenderness here originates from a situation of double contrast: on the one hand is the frailty of the child, the child's being powerless, entrusting itself to protection; on the other hand is the self-sufficiency of the observer, who can grant or deny such a protection. At the same time, however, together with the child's appearing frail and the adult's appearing self-sufficient is the perception of the future potentialities of the child's being, of what the child can still become and the adult can no longer. Tenderness crosses through this complex intertwining of sentiments, in which there lie jointly a contrast and a reciprocal reference of appearing and being, frailty and potentialities, what is powerless and what is self-sufficient, instant and time in its complex unfolding of past, present, and future. Only the adult can feel tenderness though, because tenderness manifests itself only through an act that measures the frailty that we are. In this way, one can also perceive its strength, for one measures frailty on the basis of the double register of the consistency that, as adults, we are and of the even greater potentiality that opens up for the child. These layers of reading are not successive but rather simultaneous; they are intuited, as it were, by contrast.

Tenderness is an ontological sentiment because it unfolds being [*l'essere*] in the multiplicity of its layers, and thereby attains being's depth. It measures, at the same time and in their contrast, caducity and persistence in time, frailty and consistency of things. In this sense, it is an ambivalent sentiment, like nostalgia, especially in its German rendering: Nostalgia (*Sehnsuch*) is desire searching for (*sucht nach*) something; thus it contains a push spurring it toward the future. In this search forward, however, nostalgia turns itself backward, toward the past, because precisely there it finds a piece of a future that has been forgotten and should not go lost. So, too, does tenderness, because its gesture glimpses the child as already adult; it glimpses what the child loses and gains; it sees how frail, unstable, and great that which the child is right now is; and it wishes that nothing were to go lost and everything could reach its fulfillment. Those who perceive all this are the ones who have already lost, and mislaid in a repressed past many of the possibilities that they now see lie in front of the child. That they know how to recognize [such possibilities], however, is because they have not properly lost them. Ultimately, in the child they recognize themselves. They recognize that which was and that which could return to being; they recognize that which is as a possibility that is neither exhausted nor forgotten.

In tenderness lies the strength of letting be that which is, but which is also threatened of existing no longer. To paraphrase Nietzsche, there is solicitude and passion for what is transient. For this, one needs overabundance and measure at the same time: the overabundance of one who looks at things and goes beyond them, and the measure of one who, by going beyond, carries with oneself inexorably, as a burden and a commitment, that which one transcends. Tenderness quickens its pace and does not stop. Yet its quickening takes the form of lingering: the reason for hastening is that one must keep pace (and here are the very peculiar measure and lingering that characterize tenderness) with that which is transient and could disappear at any moment, but the origin of hastening (and here is the overabundance) is not to let oneself be overwhelmed by transiency and thereby lose that which is worth being and remaining.

Tenderness is a form of ethics, not because it describes a virtue, but rather because it gathers the moment in which the ethical and the theoretical decisions reconnect. Retrieving the present as the time of existence has led us to a reconsideration of the tasks of education and to the discovery of new virtues. The theoretical path discloses new practical horizons. Such horizons, when rethought in their radicalness, allow for new theoretical acquisitions, and moreover enable one to understand the relation between theoretical and practical dimensions in an alternatively meaningful connection. On the practical side, tenderness corresponds to what wonder is on the theoretical side. Neither is a form of activity; rather, they are modes of evidence. Being transpires (as it does in the theological mode of the ontological proof where, as it will be seen later, God's being proclaims its own inescapable existence). And being transpires as the initialness that is to be approached, on whose pace one measures one's own, on whose fire one exercises one's sight, on whose rhythm one stretches one's own existence. Tenderness appears then at the bottom (at the end, but also as the basis) of those ethical virtues I have tried to suggest. Tenderness is beyond virtues, yet it is not properly beyond ethics (let alone an ethics of the present) because it is never on the other side of it. It is at a point of intertwining and overturning, where evidences touch points that have to do with the origin.

There is one last reason for tenderness, and this is something that intimately belongs to the philosophical work, as I understand it. Philosophers feel wonder because in their speaking they always say a word that is the first. The philosophical word always says the beginning; it discloses the world for the first time and unfolds the origin. As Merleau-Ponty states, the philosophical word is like "the child's"[7] expression. In this, philosophy is like art, as Merleau-Ponty himself recognizes, yet with a peculiar break because philosophy inaugurates the world not through a creating gesture but through an act of reflection—a second, reflected gesture. [Philosophy] says the beginning starting from a condition that is outside the beginning. With an unheard-of presumption, it even asserts that the only possibility of saying the beginning, a beginning that is forever lost and yet told anew, delineates itself precisely from out of such an unbridgeable distance. [Philosophy] says the present because the present, which is not the eternal, is a beginning that has then already begun, that will end, and that has nevertheless constituted something forever.

This gap, which intimately runs through philosophical discourse, always on the verge of silence, is the most hidden root of tenderness; it is also the point of intersection between philosophy and poetry. It is such an interior diffraction that one should protect; that is, that one should bring to words without thereby filling it. The philosophical word says the beginning again; therefore, it says the principle or the origin only as that which has always already occurred—as something to be reflected, rediscovered. [The philosophical word] is not absolute but instead derivative. Yet, the beginning that it says is truly the new absolute, because it discloses a world that was not there before. Without being absolute, [such a word] is then nevertheless constitutive. The beginning that it says is not the origin. Yet such a word is, in this way, a way of saying something about the origin, a lateral way of saying that does not possess. It says [the origin] as a precious good that needs to be protected but over which there is no ownership, as that which is mine to the highest degree and yet does not belong to me.

How can one look at such an intimate break, for which my innermost I does not belong to me, if not with trepidation—with tenderness? How not to tremble for this further move when, having secured the present, one must narrate it, keeping it from becoming history (a simple legacy of the past) or a dream (a fleeting desire for the future)? When one must share with others that non-me that is still me? When one must make room, in time, for other presents? Storytelling [*il racconto*] then becomes a difficult yet necessary test for a philosophy of the present. First of all it will be, as usual, a journey of the self toward itself finally to become the story of the unfolding of such a journey. That is, it is the tale of a subject that is not the subjectivity of the I but rather the coming to evidence of the temporality of the subject and, therefore, justifiably so, also the evincing of the intertwining of other presents, the exit from isolation and solipsism. All this, however, without dismissing the oblique, transversal mode, which is the only possible mode we have to say alterity—all alterities, including the alterity of the true.

CHAPTER FOUR

Tale without Author

THE TALE OF PHILOSOPHY

How can we tell the present? Even more radically, how can we think of the present? Has not philosophy, because of its very structure, always thought through categories of the eternal, or in any event outside of time, or by essentializing time under the form of its concept?

One could probably debate for a long time, and then perhaps see that abstract thinking contains a wise and passionate dramaturgy; that concepts are not only the result of labor, but they also include life, history, and time. Behind the fight of argumentations and the armor of logical structures, one would glimpse the pulsing of life, a pulsing that, as it should be, is kept in check and at bay. The tale format is not foreign to philosophy, if by "tale" one does not simply understand fantastic invention. Philosophy also tells tales. It tells tales after having reduced things to the essential (or at least to what it believes to be the essential). Once it has grasped it, [philosophy] retells [the essential] in a myriad of ways, in the most varied forms, and attracts toward it not by being satisfied with simply moving feelings, but also by involving reason.

In this respect, philosophy is ambitious and wants everything. It wants to reconstruct a world after having deconstructed it; it wants human beings to live in it in their entirety, with their doubts, prudence, and leading reasons. A bit ironically but ultimately also with seriousness, do we not refer to "the great tales of philosophy" when we wish to indicate the fundamental systems, the basic options, the dramaturgical scenarios marking the history of our thought?

In any event, without getting lost in too detailed a recognition, I think that, within philosophy, a narrative thread can be identified precisely where [philosophy] engages the supreme ontological (and theological) questions. As a theoretical proposal too, I would like to follow such a thread so as to return philosophy to the art of storytelling, which I consider to be the confession of one's own immanent temporality. In his 1928 courses at the *Collège de France*, the French psychologist Pierre Janet perfectly highlighted the knot keeping together present and narration, present and memory: "What characterizes the present is not the preparation of memory, it is precisely its narration, it is narration. The present is a narration made precisely in the moment in which we act."[1] The present then, or at least the present to which we intend to refer, is the expansion that has been achieved not by exiting temporality but rather through a conscious exercise in temporality, such as the one found within narration. This also shows that the emphasis on the present is not alternative to the reference to memory, because it is rather an enactment of the latter, a sort of *distensio memoriae* [distention of memory]: "The present requires phenomena of memory. There must be memory so that there may be present."

Dwelling on tales should therefore not appear as a detour nor as, an albeit temporary, relinquishment of the theme of the present; rather, it should be considered as an enrichment and specification of such a theme. If we wanted, we could call it a lingering. Looking for reasons to linger, suspending the flux of the main theme, and then gathering, at the end, the various trends of narration that highlight the wonderful, secret correspondences—is this not perhaps congenial to storytelling? And dwelling upon a thickening, of which lingering may be a form—is this not congenial to the present as threshold? Beside all analogies, which might be extrinsic, I see in the narrative character of philosophy a way of complying with the temporality that runs through it. I see in the author of the tale, in the one who weaves the threads of the narration, a subject capable of instituting a present (or, vice versa, I see an inexhaustible fund of possible narrativeness in the one who institutes a present). Furthermore, this route through narrativeness will lead us back in time, to a great classical text—that is, to the place that I have evoked more than once: the formulation of the ontological argument. We will thus encounter not only a fundamental text, but also one whose appearance is fundamentally nonnarrative. From this perspective, too, the ontological proof constitutes an extreme—the extreme of a narrativity without author. The in-between in which we are is defined precisely in the collision with extremes, such as the *cogito* and the ontological argument, with respect to which [the in-between] defines itself (by detaching itself from, and returning to them). When one moves from the needs of modernity, from its hopes and failures, but especially from its discomforts, one is always induced to return to those fixed points, which have sedimented in the classics, from which everything has started and perhaps starts again. We will thus see that unsuspected paths and developments occur within such a field (which does not institute a secularizing linearity of the kind: from the ontological argument to the *cogito*), that is, within the extremes of the emphasis on, and the irrelevance of, the narrative author ([extremes] that are mirrored in the emphasis on, and the irrelevance of, the ontological author). After the various crisis and even tragedies of the I, Celan's proposal, on which I will reflect later, delineates a new starting point. It is [a starting point] that, by making us assume the present that we are, works toward its extension and reconstruction, without thereby denying or eluding the feature of break in history that the I of modernity has interiorized. But I have anticipated even too much. For now, let us still linger shortly on the features of the tale, and let us be inspired by Anselm's argument.

TALES AND *DISTENSIO TEMPORIS*

Several [scholars] have already remarked how, as fables attest, in narrations the true interest is not in the plot, in the story, but in how the story, which is already known, is told. In addition to the even maniacal reassurance that children wish to attain through an always identical repetition of the same, a fundamental need to hand down is present in the very structure of narrating. What is handed down comes from the past, or is already known; nevertheless, one wishes to repeat it in the present, to turn it into a possible present. It has also been remarked how there is an affinity between the activities of narrating and translating. The affinity lies precisely in the fact that in both cases the text already exists, the content is given; one just has to retell it. This is possible because in both instances the text is not a codified model, but rather an *Urtext*, an origin that cannot be identified in an original model; it is the trace of an absent that one wishes to be present. The tale translates from an original that has disappeared; it retells an already said that nevertheless is resaid every time as if for the first time insofar as it always and only happens there, that is, in the act of narrating. The tale has no other function than to make its content present in the specific moment—and thus, in some way forever. Not different is the case of the translation, which also makes present and accessible, as if by miracle, a text made of inaccessible signs, the meaning of which was precluded (this is evident when confronted with a language that is for most of us truly unknown such as, for example, Chinese or Japanese, the signs of which appear illegible to us). The translation does not translate from such a text, though, or it would fail even more than it already does. Insofar as it is the original, the written text is perfect, and all changes can only worsen it. Yet such an original also has an origin from which it comes, a non-text from which it descends and to which it hints. Through such an allusion, the translation searches for the original non-text and translates from there. It is a sibling, not a child of the first linguistic original. Therefore, like the tale, [the translation] too retells, and it actualizes like the tale does. Translations and tales tell the name, bring to speech a name that has never been said and, in human speech, restore the holy of language.[2]

Tales venture in the dimension of time. Even more precisely, they unfold time, because they transcribe the original into the rhythm of words and pauses. The function of tales is to make such an original that has disappeared last, and to give it the breath of the present. Ultimately, tales can live on one idea alone: that in the end the good always wins, which is like saying that it exists. Yet, tales do not utter; they only exhibit such an idea. Like in a nursery rhyme, the good takes shape through progressive extensions, and it progressively encounters obstacles, even defeats; however, in the end—and there too—not suddenly, as if by a miracle, but after more than a trial and fatigue, [the good] goes back where it started, but now it is self-confirmed, and even richer. So much useless fatigue [is employed] to tell what one already knew and even expected since the beginning (like in the happy ending of commercial movies and novels). This is because the interest is not for the matter but for the how (and here is the difference from the happy endings of trivial stories, where the happy end [in English in the original] contradictorily falls apart in the end so as to solve the plot: it brings to an end, it does not give consistency to reality). One is interested not in what will happen (one knows that since the beginning) but in how it will happen, through what adventures, which are the real content of tales. Tales are different from chronicles not because of the material they contain, nor for the subjective aspects that are made to intervene, but rather because in the spires of telling one searches for the essential and tries to hold onto it.

Not even in this way, however, has enough been said; as if the prevailing interest were in the form, which is itself inessential to tales. The most proper interest keeps residing in the act of narrating, and precisely in the time of tales. The narrator unfolds time, I have said. Here the *distensio temporis* unfolds in the sense of an objective genitive: the narrator's sovereignty determines the length or brevity of the events, marks interruptions and repeats. In tales, however, there is also a *distensio temporis* in which the genitive is subjective: time imposes its rules—not so slowly that tales coincide with that sort of cosmic time that is the flowing of life in its objectivity; not so quickly that they seem not appropriate to it; or, perhaps even more slowly, as if in a slow-motion shot, but through an isolation of the particular that symbolically projects it in the form of the whole, and then (why not?) so surprisingly quickly as to be an explicit contestation of life, a magic white page interposed with reality. In any event, the freedom of unfolding time, [freedom] that belongs to the narrator, delineates itself in relation to the need of the self-unfolding of time that is related to life.

In all cases, tales unfold within these two limits; therefore, they have to do with time. Between the time of pure interiority and the time of indifference, however, they trace yet another time: the time that precisely the present is. In tales, the present is, if I may say so, the time of time, the secret soul of the *distensio temporis*.

As long as the narrator speaks, the good exists, the positive hero lives (or, like for Shahrazad, life is prolonged).[3] Trials and defeats tell the protagonist's existence in the present. In retelling, in telling an old and already known truth, tales make such truth exist in the present; they attest—that is, they testify that existence within time and within the being-present of time necessarily belongs to the nature of such truth. Thus the narrator does something similar to the translator, who does not translate a text into another—the original text into the text of the translation—but rather attests that even the originary non-text, which is pure essence, presently has a temporal existence.

Retelling is not to repeat but rather to step back a great deal, return to an origin, and stretch out toward an announcement. That is, [retelling] is to turn an essence into an existence. It is to return to the essence, to the essence that is lost at the origin but is hidden in the name. It is to follow the thread, the trace that has gone astray, and to land on the originary essence, as if by a miracle. One cannot stay by the essence, though. One must redescend, and form new names and traces. One must thereby generate new existences, which simultaneously remember and forget the essence of the origin. It is not a matter of prolonging but of overturning the one into the other (the essence into the existence) by going through contradiction and negation, overturning and reversing. Time grafts onto the eternal, negates, and at the same time actualizes it. Like philosophy, tales are a relocation within the world. For both, the material is always the same; it is simply a matter of rearranging it again. This constitutes not a formal but a substantial operation. In the end, through a tale, through an unfolding of time in the present, a world is born.

TALE WITHOUT AUTHOR

If tales are like this, then they need a philosopher to unfold them, or at least an author who assumes the responsibility of having molded time in a specific manner. On the contrary, the great principles of Western philosophy and theology seem to follow a different logic. They do not seem to deal with time but with the eternal, and an atemporal objectivity displays itself in them. If one really wanted to consider them as tales, they would be tales without author.

Let us reexamine Anselm's argument from this perspective: "In my judgment, neither this work [the *Proslogion*] nor the other, which I mentioned above [the *Monologion*], deserved to be called a book, or to bear the name of an author . . . I accordingly gave each a title."[4] A title is useful to give a name to the tale, but the tale does not deserve the name of the author. The true is ancient, and appearing like a conceited innovator would not be good. In truth, that which must be said has already been written.

Inquiries are born out of another's request. It is others who insistently exhort one to meditation. When one devotes oneself to meditation, the effort seems to bring no results. Rather, it distresses. Despairing in the ability to achieve some result, the most opportune proposal seems then to give up and chase all thoughts away. Precisely then, though, does the argument one was seeking for come as if to impose itself, despite the fact that by now one only wishes to resist it.

When confronted with such a refined dramatization, the suspicion that this too may be a tale, albeit without an author, becomes licit. What one seeks—the object of proof—escapes, so that one despairs of it; and the searching I almost withdraws. It is precisely then that what was deemed unachievable imposes itself in a sort of objective unfolding; conversely, the one who is searching balks thereby renewing the conflict at a different level.

Such are, in an intertwining, the lines of the introductions with which Anselm prefaces the double and converging essay bearing the name of *Monologion* and *Proslogion*. One could bring many reasons for such a prudent and reserved attitude. All of them, which would relate to the culture of the times, would help to cast light on a certain epoch and customs in which, one could say, the subject was not yet (philosophically speaking) born, and the truth had not yet coincided with the new. Yet [such reasons] would not suffice because they would not touch "the matter" but only the form of its manifestation. There is something more in this tale that does not perceive of itself as tale, in this author who does not name himself. There is the fact that we are at the limit of all possible saying. We are where the saying—which is always a tale—borders on that which cannot be said; but we are also where saying the impossible accounts for that which on the contrary must be said (through other tales and within time).

Anselm's text is a splendid document—among the greatest that were ever written—on God's nature. Existence also necessarily belongs to such nature. Whereas the *Monologion* may still suggest that one may come to God, the *Proslogion* is rather a sinking into God. God cannot be properly demonstrated. Yet, he is already there. When the fool names him so as to deny him, the fool does not know what he or she is talking about, because he or she presumes to be dealing with an entity whatsoever—a chimera, a blessed island, a half-human-half-beast monster—which can be named yet to which one can deny existence. The fool thinks that it is simply a name, a representation of an excited mind, an unbounded desire. God's name is not such. Once one encounters [God], one can no longer get rid of him. One can certainly elude him, put him on a side, or fight him. But one can no longer deny him. If one questions him, then God wraps one up in his depths. At the end of a path of reason, that-than-which-a greater-cannot-be-thought also appears as the living, the wise, the blessed, the eternal and, in a word, the compendium—the cipher for "whatever it is better to be than not to be."[5] In a new synthesis and with a new development, he is even a being "greater than can be conceived."[6] He has such a nature that nothing can be taken away from it or he would no longer be the greater being of all, because the power of one's thought could then also think the very same nature yet with the addition of that which has been denied to it. Nor can anything (be it a desire, a good, or a splendor) be added to it because if that were the case, once again one could think a nature greater than the highest. Existence is implied among such spires of thought, because taking it away or adding it as if it did not belong to God's very nature would be tantamount to admitting that God is not [God].

It has been said the Anselm's prevailing interest is logic. When compared with that of master Augustine, Anselm's style (although at times, at least in the letters, overabundant and even excessive as for its *pathos*) certainly appears as less existentially involving. One can perceive the presencing of a passion for pure argumentation, which also translates into a sovereign automatism in terms of unfolding the issues that are addressed time after time: first is their sectioning, followed by their decomposition, and then their separate confutation starting with those questions that appear the least sustainable. Yet, when it is not pure formalism, even logic contains a tale. Argumentations disjoin that which, when it is united, does not appear as discernible or join that which, when it is disjoint, appears as disconnected. We have seen this—from God as that-than-which-a greater-cannot-be-thought to God as the greatest that can be thought. A journey has been made from the double negativity of the beginning to the impossibility of a greater positive—a vicissitude around the infinite.

Such a journey marks, within the temporal succession of language, the determinations of God's nature. It narrates God's attributes; it attests God's existence. It is a philosophical journey, and thus it is entirely devoted to "the matter"; therefore, it has no author. The name that matters is the name of the matter.

THE ONTOLOGICAL PROOF: EXISTENCE THAT HAS ALWAYS ALREADY BEEN THERE

The history of the ontological proof has already been written with great precision and erudition.[7] I think that all of its possible variations have also been tried and explored. I am convinced that nothing can be added in such directions. Instead, it is better to return to Anselm and to his first saying as it has been fixed in the two titles of the decisive paragraphs, "That God Truly Exists" and "That God Cannot Be Thought Not to Exist." As is clear, at the center of attention is existence. One would be mistaken, though, if one were to think that in Anselm's argumentation there might be an apologetic or controversial intent. Anselm's interlocutor dwells in monasteries and is a believer; otherwise, he is not a real interlocutor; likewise the fool in the Psalm, who in fact belongs to a same religious tradition but thinks he can detach himself from it. Ultimately, the fool plays a game with thinking because, anticipating a split that will become actual only in future centuries, he separates word from thing. In this sense the fool is a borderline figure, a mental experiment that has value not as an interlocutor to be convinced but rather as an a fortiori proof of the goodness of the ***unum argumentum*** [unique argument].

What constitutes the discovery of the proof from an Anselmian perspective? Although it is indubitable that the attention is on the question of existence, we should not make of this one single thing with the claim "God exists." Such a formulation is in fact not reproduced as the title of the corresponding chapter. Where it appears, it occurs with the essential addition of "truly." In the subsequent chapter, what it proposes is an essential inherence in the cogent form of a double negation—God cannot be thought not to exist. The impression one derives is not that God's existence is the primary content of the proof, but rather, precisely as suggested in the formulation of the first of the two chapters, that such an existence, which is believed, is now truly discovered by thought; that is, faith is given an intelligence of its object.

Thought that thinks God's nature comes across God's existence. This is the believing thought (although, as we have remarked, the same can be said also for the nonbelieving thought) that neither seeks nor needs to arrive at God's existence, but rather sinks into God's nature, and grasps existence itself precisely as God's most certain and in fact originary belonging. What echoes in it, as if in a counter-melody, is the evangelical exclamation that "Truly this was the son of God!" (Matt. 27:54) and, in an even more punctual manner, Luke's tale of Emmaus, when the unknown [person] reproaches the disciples for their being foolish indeed for not understanding what is in front of their eyes. Having explained the Scriptures, [the person] arouses such a desire in them that they themselves, the foolish disciples, ask him to stay, to become their guest, and to share their dinner. In the repetition of the Lord's gestures, he then reveals himself as the one who "has truly risen" (Luke 24:25–35), according to the word of the eleven who are gathered.

The core of Anselm's saying is perfectly analogous to that of the two evangelical quotations. It lies in the "truly," which has no logical-juridical character but rather hints to the surprise of the finding. Like the child who, in Elizabeth's womb, rejoices in the encounter with the Lord's mother, existence is not produced in the proof nor is it achieved by the proof; rather, it is discovered, that is, found again. It is truly as I already knew that it would be.

The subsequent move is no less surprising because it attests that it cannot not be this way. The form of the double negative institutes a necessity that is at the same time more and less cogent than positive necessity. It is more cogent because it does not admit of any exclusion, limitation, or reservation. It is less so because it does not assert a positive cogency, as if nature and existence co-belonged, but only an insurmountable condition—when one thinks of God one cannot lacerate the nexus of essence and existence.

The true discovery then is not that God exists, but that God truly exists and cannot be thought as not existing.[8] In God, existence has always already been there [*è già da sempre là*] in his nature—just like and unlike in human beings. Like in human beings, and in the finite, because here too existence is always already there [*è già sempre là*], and thought can only begin with an existence that has always already been there. Unlike in the finite, because in [the finite] the anteriority of existence does not coincide with the nature of the essence; rather, it only belongs to [the essence] accidentally. Death insistently threatens it, and increasing and decreasing, wearing down and enriching belong to it. God's existence is the same and different. Nevertheless, it is irrevocably there, the seal of an anteriority that is nature; it is at the origin of the true. Transposed into the language of temporality, the present of God's existence truly coincides with God's eternal nature. Like and unlike in human existence, whose self-presence is punctual and threatened.

"Then they told what had happened on the road, and how he was known to them in the breaking of the bread" (Luke 24:35)—thus the event at Emmaus ends. The existence that has been recognized narrates the gestures of the recognition. Philosophy also begins this way, by transposing the image. The ontological argument has enabled us to encounter the extreme point: an anteriority that has always already been there; a true that is ancient because it is anterior, always already there; an unmeasured measure; a nonposited limit yet an encountered barrier. Thinking takes its move from this. Not even a thought that wishes to say the beginning is capable of catching the beginning in its own beginning. We always think *post festum*, after the beginning—even when we go back and look for [such a beginning], even when, like Anselm, we feel that we are situated in the beginning that is truly first, and we perceive all its joy. This is so true that, as we have seen, even in this case of so little narrative appearance, even in this case that seems to deal more with the eternal than with time, we can only narrate. That is, through words, signs, gestures, in the disconnected temporality of succession we unfold the surprise of being able to discover such an ancient truth today; we unfold the surprise of encountering today (that is, after the beginning, after the forgetfulness of the beginning) a still and yet again unexpected anteriority, an anteriority so unexpected that it presents itself as new.

CHAPTER FIVE

The Tale of the I

THE TALE OF THE I

Not all philosophical tales are without an author. Wearing a dressing gown, warmed by the fire, surrounded by familiar objects (paper, wax), well protected in a room outside of which the world is reduced to hats and cloaks seen from behind the windows and under which automata might be hiding, the I makes its appearance as author. One cannot deny that its scenic presence is that of a great actor. As one might say, it takes up the scene. It towers alone, and settles the entire tale within itself, within its own vicissitudes.

There are many correspondences with Anselm. Here, too, the same ontological proof occurs, and even the style, transposed from a monastery to a bourgeois house, is the same—a spiritual meditation that, day after day, retraces the days of creation. Whereas in Anselm one is totally immersed in the highest thing, in Descartes, with a 180-degree translation, everything occurs within the subject. The world is no longer that of archetypical ideas, which in God precede creation and give rise to it; rather, it is the modern and pale world of mental representations. Yet, within such a faint arch, the same modality and in the end the same question propose themselves. In the divine essence, Anselm had found the existence that has-always-already-been-there, the necessary co-belonging of the one and the other. Descartes is looking for the same thing, he seeks the same coincidence and co-belonging. Yet he does not seek it from out of the immense object, from out of a thought that one seeks at the same time as one finds refuge in it, into which one runs and which ultimately imposes itself. Rather, he seeks it in the residual object that the I is, at the end of a mental and spiritual experiment that revokes the certainty of everything so as to start all over anew with the minimum—the presupposition of an I that cannot be eliminated. The existence that is always already there is the I's finite existence. This existence is immediately also the I's essence. What am I, I who exist? I am a thinking thing, that is, a thing that doubts, intends, affirms, denies, wills, does not will, even imagines, and senses.

The indubitability of the I is the indubitability of the fact that I cannot think of myself as not existing. For were I even to think of myself as [nonexisting], in the act of thinking [such a thought] I would exist. As one can tell, this too is an ontological proof, albeit turned upside down in comparison with the earlier one. This [proof] does not attest an existence nothing greater of which can be thought; rather, it certifies that the I is that nothing less of which can be thought, and that for this reason it is that which necessarily exists. The argument is the same, even when its application is directed toward the opposite extreme of the ontological ladder—toward the minimum point of coincidence of essence and existence that is constituted by the *cogito*.

When reduced to its minimum, the ontological proof shows its own validity even with respect to the *cogito*, because at this level also it attests the co-belonging of essence and existence. The co-belonging is only punctual and precarious; it is in the act of thinking. It is a co-belonging surrounded by doubt and not by joy. It is a co-belonging that isolates. Nevertheless, it is a fixed point with which one can hope to lift the world, like Archimedes. Naturally here, unlike in Anselm, what towers is not the matter but rather the subject. In the tale that is being told, the author is indeed the only certainty. Yet confidence is not lacking that a new tale delineating the world might unfold from out of the author's world.

The Cartesian scheme is certainly not void of ambition, given that in the fifth meditation it culminates with the ontological proof properly named; and there, too, having turned the hemispheres upside down,[1] it ascertains the same necessary co-belonging of existence and essence. Taking up Anselm, it recognizes existence as necessarily belonging to God's nature, because the infinite in act cannot but also exist actually. If it exists, the triangle is the figure the sum of whose internal angles equals 180 degrees; if they exist, mountain and valley describe the same and indissoluble geographical reality albeit seen from two different perspectives. Likewise, if God is possible (to anticipate Leibniz's development), God is the being whose essence implies his existence. As one will notice, the exemplification is not by correspondence but rather by mirror image. Like all finite things, the triangle is certainly possible, yet it is not certain that it exists. If he is possible, God, who is the infinite being, is certainly existing. That he is possible has already been demonstrated by the third meditation, because the indubitability of the evidence of the *cogito* can only be ensured within finitude, and finitude is in some sense subsequent to the infinite. The possibility of the infinite is the condition to be able to think the finite. Thus, the infinite is possible. Therefore, it certainly (even more certainly than the finite, which is always mobile and ungraspable) exists.

As tale of the I, philosophy has had a rather odd destiny. It has not produced an attention to the subject so much as a focusing on the subject's world—on ideas as representations of the mind. In some sense, the classical seal has continued to mark the reflection. Though with a 180-degree rotation, yet firmly situated in the coincidence of essence and existence, philosophy has continued in its wish to describe the world: no longer the world of creation, however—that is, the majestic and objective nature—but rather the world granted by the certainty of the *cogito*—its ideas and representations. At the end of this path, as we know, there lies either the anatomy of the mind fragmenting even the I in the punctuality of its representations or the I as world of idealism, with its tenacious retrieval, after Kant's precarious balance, of the themes of the classical period. The latter is a patient repetition that restores the idea to its own worldly richness, and releases it from being a simple and punctual mental representation.

The tale becomes then a cosmogony because the I generates worlds, and appropriates ever new provinces to itself. Albeit upside down, the germination of modernity reproduces the efforts of the Middle Ages that, starting from the highest being, try to secure a terrain of consistency for the finite. In modernity, having anchored oneself to the principle of the minimum and setting out from it, one seeks a new beginning, and one dilates it until it discloses the whole. From Descartes' modern to Hegel's modernity the path is long, rich in adventures, in new conquests and discoveries, in conversions and surprises. Yet it is linear, guided by the omnivorous augmentation of the one who wishes to close the circle and restore to its dignity the upside down ontological proof constituted by the *cogito*.

The author of this new tale assumes less and less defined outlines, and is disquieting because of [the author's] own very expansion. [The author] is no longer protected by the warmth of the fire and the everydayness of a dressing gown; it becomes spectator and judge of the course of time, an impersonal eye within an absolute history. Foucault grasps this with great finesse. The classical age of reason quickly archives the subject because it has simply used [the subject] to bring a little shock to the world. Void of the subject, there remains representation as such—prodigy of a representation that looks at its own spectators. The king is absent; only his family is present. The room is background and not place; yet, on the threshold, in the background, there is a character that looks. The canvas is without the painter. A sort of picture overflows its frame and invades the world. Representation represents itself and becomes the whole, filling it with its own absence. The author is lost because of the very height to which it has elevated itself.[2] The principle of the minimum has overturned itself into the principle of the whole. In this way, though, as one cannot fail to notice, everything is put at stake—everything, including Descartes' indubitable I, can be doubted.[3]

The existence that is always already there, that clings to the essence that it itself necessarily is, the trepid but certain beginning [*inizio*] becomes mere commencement [*cominciamento*], that which always and only needs indeed to be forgotten. One certainly no longer discovers—with joy—the necessary coincidence of essence and existence in the highest being. Nor does one cling with intrepid stubbornness to a punctual existence in which the essence too is given. Everything stretches in absolute coincidence, which only waits to become known in order to be perfect. In order to know, however, one needs to forget. It is no longer [the case of] reminiscing about a truth that lies behind us, of imitation or asceticism. It is a truth that runs in front of us, always a step ahead of consciousness, which finds itself precisely by forgetting itself.

Freedom does not have fathers, mothers or genealogy. It spurs ahead, it calls for incorporation, where remembering is precisely interiorizing, that is, destroying and forgetting that which was and remained outside of us, external/extraneous. Because it has repudiated the beginning, the tale has lost its author on the way. What matters is the unfolding, and this occurs by itself, with no guide, no controls, as sovereignly absolute. The tale is worth being lived, but its author no longer inhabits it.

One day the big *O* with which the *Ottocento* [the nineteenth century] begins exploded, and philosophy as the great tale of totality started to be abandoned. The age of specializations began for it, too, after the relinquishment of the myths of science, progress, history, humanity, consciousness, or dialectics.

We still need a great tale, though, a tale whose possible author remains unknown. Too many denials, failures, and illusions divide us from Anselm's clarity, from Descartes' certainty, from Hegel's ambitions. Too great a desire and need separate us from the silence of the various specializations.

Let us return to narrating. What? A tale without author? An author without a tale? A sense without meanings? Meanings that have no sense? As in Baal-Shem's tale,[4] one can return to the same place in the woods even when one has lost many things; even when one has forgotten formulas, rites, and gestures. One can at least set out on a trip backward, toward those beginnings—which have always been double: essence containing existence as that which has always already been there, and always-already-there existence containing essence at least for an instant. If the circle could be closed, then the universe would return to being in order, meanings would have their sense, and sense would contain meanings.

Memory would no longer be a journey outside itself so as to lose itself, nor would it be the impossible labor of appropriating and encompassing everything so as better to help the process of consciousness. Where is the place of such an experience, though? And how can we live and narrate it?

THE I AS A ME

The path has been indicated not by a philosopher but rather by a poet,[5] who is in the train of other poets and even of literature. This poet writes a tale, that is, the real tale of a stroll that is also the tale of a real stroll of another poet, such as a writer has narrated it; and in this way it is perhaps the beginning of a tale that is simply literary. On January 20, while going through the mountains, Lenz (Büchner's Lenz, the real Lenz)—that is, not the artist who debates art related issues but rather He *qua* I—makes the extraordinary experience of encountering himself as a self-estranged I.[6]

The experience of January 20, from which time begins again, is very personal and solitary. Surrounded by the mountains, the I encounters itself as a self. There begins the journey that will end up in the streets of Moscow, where Lenz was found dead. On January 20 (1942), the final solution too was decided, the tragedy that Celan escaped after losing all of his family and also the joy of going on living. It is a cursed date. Yet it is the date of an encounter, albeit distorted and upside down—the I and its estrangement; estrangement and, at its end, the I.

The problem had been posed. No longer God or the I, nor the still peaceful alternative between the pair I-you and the pair I-it; instead, the radical question: the I and its own self-estrangement from itself; estrangement assuming the power of a subject, of an I in front of an I. The subject is confronted neither with certainty nor with its own freedom; rather, it is placed in front of its own self yet as capsized into estrangement. It is a place of perdition, of destruction rather than of salvation. Yet a finding [*ritrovamento*] also (perhaps) occurs there. Like a line disrupting the chain of automatic developments, of the triumphal pretensions of history, it is "an act of freedom. It is a step."[7] What matters is the finding, regardless of how it may have happened, even if it happened on the path toward death. The I as another; yet, also and jointly, the other as an I.

As we know, encounters fail most of the times. The conversation in the mountains, which Celan imagines with Adorno, does not succeed, cannot give birth to a true dialogue, cannot reconcile one with nature. It does not even succeed in maintaining silence, which Adorno has imposed on poetry after Auschwitz. Yet the encounter has occurred.

> And I know, I know, cousin, I know I've met you here, and we talked, a lot, and those folds there, you know they are not for men,

> and not for us [here is the encounter under the sign of negation] who [however] went off and met here, under the star, we the Jews who came like Lenz [like Lenz] through the mountains, you Gross and me Klein, you, the windbag, and me, the windbag, with our sticks, with our unpronounceable names [the name too is under the sign of unspeakability], with our shadows, our own and not our own, you here and me here—"me here, me, who can tell you all this, could have [the failed possibility of the encounter] and don't and didn't tell you [the failed encounter indeed]; me with a turk's-cap lily on my left, me with corn-salad, me with my burned candle [the Sabbath candle is loved and welcome again precisely because that tradition has been consumed as consumed is the time of existence], me with the day, me with the days, me here and there, me, maybe accompanied—now—by the love of those I didn't love, me on the way to myself, up here."[8]

The rose belongs to no one;[9] that is, the name that gives a name to the rose is no one. Not no name; rather, not a someone: no one, the non-beloved. Yet no one *here*: that is, I on the way to myself. It is neither any way nor simply my way; yet it is not someone's way. It is the way of *no one*: no one, who is an I searching for the me. The encounter is the counterfigure of Diogenes who searches for himself. Here, too, with a candle, albeit burnt to its end. And the search is not for God (a God whom we have killed), in a Nietzschean manner. The search is for another name of God. In the almond, like an empty frame giving eyes back their cut, there is nothingness, a royal blue nothingness that persists (*steht und steht*).[10]

After Descartes and after his arc in Nietzsche (Löwith's *God Man and World from Descartes to Nietzsche*),[11] an encounter where the I is the stranger whose name is no one and where God is nothingness opens its way.

This is not a destructive nothingness, though, and even less is it a nothingness attained through subtractions. It is rather like a suspension: "*Von deinem Gott war die Rede, ich sprach / gegen ihn, ich / ließ das Herz, das ich hatte, / hoffen: / auf / sein höchstes, umröcheltes, sein / haderndes Wort—// Den Aug sah mir zu, sah hinweg, / dein Mund / sprach sich dem Aug zu, ich hörte: // Wir / wissen ja nicht, weißt du, / wir / wissen ja nicht, / was / gilt.*"[12]

In order to begin, all this is very little. It is a thin ice sheet. Yet, within the constraints of the times, it might even be enough. Perhaps it is also an exit from the pattern of the I and God, the subject and the object, and the infinite reversals. Something like a displacement occurs. Zeroed by negativity, where oblivion and memory meet, nested like wine in shells, sunk in the primordial darkness of sexuality, another encounter happens. And there is an announcement that covers itself with salvation, or at least with impatience: "*Mein Aug steigt hinab zum Geschlecht der geliebten: / wir sehen uns an, / wir sagen uns Dunkles, / wir lieben einander wie Mohn und Gedächtnis, / wir schlafen wie Wein in den Muscheln, / wie das Meer im Blutstrahl des Mondes. / Wir stehen umschlungen im fenster, sie sehen uns zu von der Straße: / es ist Zeit, daß man wiß! / Es ist Zeit, daß der Stein sich zu blühen bequemt, / daß der Unrast ein Herz schlägt. / Es ist Zeit, das es Zeit wird. // Es ist Zeit.*"[13]

Even here, the ancient format of the ontological proof returns, in an umpteenth, dangerous variant: not God's stable and necessary coincidence, not the I's punctual certainty, but rather the I's identity with its own estrangement, in an encounter that simultaneously attests the existence of the I (who exists) and its essence (which is stranger to itself). From out of this place—it is time, it is time that it is time—from out of this site lost in the woods, in which words themselves have become silent and gestures and sainthood have been forgotten, it becomes once again possible that impatience makes the heart beat faster. Even while one utters words that are only negative, obscure, disconnected, obscene, one can perhaps even let hope run free. Because, ultimately, a thin, impalpable, imaginary yet earthly thread—a meridian—may suffice to travel the world.

Descartes, the man of philosophical evidence, one day used a curious metaphor: with respect to thought one should act like a knight with a horse, that is, a knight who loosens the reins and then immediately pulls them toward himself, gives the animal free play and withholds it, incites and recalls it. Balance is excesses that wisely follow one upon the other—thus seems to suggest such an enigmatic wise man.

THE VICISSITUDES OF THE I

"Ladies and gentlemen," says Celan sixteen times in his brief speech. As those who profess thought always do, he speaks to an audience. He speaks *in front* of an audience. He speaks while being delimited by his interlocutor-other. One can speak only because the word resonates in a defined space and time. It comes back through the others. And the others, those who listen and misunderstand, reassure us of what we have said, although not always of what we meant to say.

Celan also speaks of others, of other poets who are dear to him—Büchner, Lenz, and their characters, Danton and Robespierre, Camille and Lucille, Leonce and Lena. For the one who speaks speaks from within a history and a tradition. No word, not even the poet's word, which is so sharp and essential, is absolute. The world does not begin with our word. [The world] merely opens itself up in its being traversed by a sign that is revelatory as well as foundational but only in a certain sense, because in no case does [the sign] by itself institute the real. Words modify other words. Through this modification, which is at times imperceptible and at times rough, one says the same while saying the other, and one says the other while repeating the same—within a circle that does not limit itself to repetition but rather designates a constellation of discourse.

What does Celan speak about? He speaks about art. Or perhaps, and better, he speaks *within* art. "Alas, art," according to Camille's expression.[14] One can place different and contrasting accents on this expression: the circumflex accent of the eternal, the grave of the historical, and the acute of the current times [*l'attualità*]. When they are applied to philosophers, these accents would perhaps mark Heidegger (the circumflex accent), who does not know any dates and rather inscribes the events of history within a granted historicity, and Adorno (the acute accent), who makes the grave of history be a burden to the point of his famous interdiction of poetizing. The acute accent concerns art, though, the timeliness of art, its dramatic reference to the now [*l'oggi*].[15] This also means the crisis and critique of art.[16]

This is so because art is here described according to two paradigms: mechanism, like in Kleist's text, who at the beginning of modernity searches for lost grace and levity by entrusting them to an ingenious as well as artificial automaton, and the Medusa's head, who fixes the instant [*l'attimo*] of life in a mortal fixity.

Art's craftiness is willing to pay any price in order to regain lost nature: "Whoever has art before his eyes and on his mind . . . has forgotten himself. Art produces a distance from the I."[17]

Yet something happens. A step, a scream, a death sentence, a January 20 (like, in the Gospels, John's tenth hour; see John 1:39), an encounter: poetry (which is the other of art, but which at the same time is nourished and intertwined with art) runs ahead of us, goes too fast, and leaves the artist aside so as to look for that very person (who says himself in art) in a determined place and time: "This person Lenz: the true Lenz . . . the Lenz who 'walked through the mountains on the 20th of January'—this person, and not the artist and the one concerned with questions about art—this person as an I. Can we know, perhaps, find the place where strangeness was present, the place where a person succeeded in setting himself free, as an—estranged—I?"[18]

"Poetry: that can signify a turn-of-breath."[19] Perhaps its path is this overturning, this inversion. Perhaps each poem remains mindful, for each of us, of such January 20. A poem always and only speaks on one's own account, "in its own, its own, individual cause."[20] Yet its hope is that precisely by doing this it can also speak in an extraneous cause, in another's cause, in a totally other cause. "Perhaps, I must now say to myself—and at this point I am making use of a well-known term—perhaps it is now possible to conceive a meeting of this 'wholly Other' and an 'other' which is not far removed, which is very near. . . . The poem wants to reach the Other, it needs this Other, it needs a vis-à-vis. It searches it out and entrusts itself to it."[21]

The attempt, which Celan desperately reiterates, is this inversion at the end of the journey. Then the quest for the other, the alien [*estranea*] and naked word, the universal of the word coincide with the-always-one-time-and-only-here-and-only-now. He is me. A he as an I. Like in the rhythms of breath, poetry incessantly recalls and retrieves itself so as to lead "from its now-no-longer back into its as-always."[22] Strangeness [*estraneità*] turns into the I, and concentration—the surprised attention already found in Malebranche, Kafka, and Benjamin—on the I who names and addresses finds in it a "thou," who "brings its otherness along."[23]

Alas, poetry, one could say. It changes the cards on the table, plays with words, and substitutes concepts without having to account for this.

Yet Celan's journey—which we have followed with trepidation (may I be allowed to say that, because it is a journey on the verge of the abyss of nothingness, where the I in the end dies suicidally)—has the features of an absolute and undeniable universality. His path searches for strangeness [*estraneità*] by means of the I. It feverishly concentrates on the I—as on a pure name, with no adjectives, dismantled—so that in an interruption, which is pause and turn, an encounter with the other may happen. As in Augustine and his concept of temporality, there is concentration and dilation. As in Benjamin, however, rather than having a reassuring arch of development, such a move has the figure of instantaneousness and overturning, and configures a sort of *Dialektik im Stillstand* [dialectics in the standstill].

What forcefully emerges is the representation of a condition of modernity where the relation with reality occurs at the end of a journey, according to the archetype of the Cartesian doubt, which suspends all certainties. It is not a continuous journey nor an adventure toward what is better nor a progressive destiny. It is rather a pause, an interruption, an overturning. As Celan says, *eine Atemwende* [a breathturn]. Breath stops, and starts again reversed. Likewise, the names are uttered in reverse (*die rückwärtsgesprochenen Namen* [the backwards-said names]):[24] Antschel becomes Celan, breath must stop in an *Atemwende*, one walks in a circle and, like in Orphism, the important moment is the turn, the *Kehre*, a sort of *Heimkehr* [home-turn] to the place of my origin. Or, like for Virgil's death in Broch, the end is when he turns around.

Why, and where, does the turn take place? It occurs only when, after so much odyssey outside, among words and things, it has reencountered the I, the he as an I. It has encountered an upside-down I. Not the first word in the world, the most trivial and closest, but rather the last. Not a certain and firm I, but rather an I traversed by things. Not an absolute and creator I, but rather an uncertain and upset I. Not a theoretical I solid on its legs, but rather an I that walks on its head upside down, reversed. Yet, as Celan writes, "Whoever walks on his head, ladies and gentlemen, whoever walks on his head has heaven beneath him as an abyss."[25] The last residue of secularization is no longer the God of Heavens, nor Being, nor Nature; rather, it is a small and frail I, who is *brisé* [fragmented] and upside down, and has heaven as an abyss under itself, because earth and heaven are by now one and the same within the sphere of existence.

Everything, in some sense, has already been said, except perhaps for the fact that Celan names the way of poetry "the path of the impossible."[26] For in the smallest and least determined form, that is, in the form of an I, poetry is, as he says, "pronouncement of the infinitude of mere mortality and futility (*Unendlichsprechung lauter Sterblichkeit und Umsonst*)."[27] [Poetry] says, names the infinite of mortality and finitude, of what is for nothing, that is, of what is not nothing but also pays nothing. Poetry walks on this mortal infinite, and searches for a thread. Celan calls it "the meridian": "something . . . abstract, yet earthly, terrestrial, something circular, which traverses both poles and returns to itself, thereby—I am happy to report—even crossing the tropics and tropes [*Tropen*]."[28] Why could we philosophers, we philosophers of this philosophy, not call it "interpretation" and say that, anchored to the finitude of the I, we have set out on a journey across the poles and the tropics (the place that takes its name after the turning of the sun), across the entire globe in order to say such an infinite mortality that is for nothing?

THE COORDINATES OF THE I'S JOURNEY

Let us now try to fix the coordinates of the journey as if on a nautical map, and let us do so by placing in order, in a chronological order, the points on which we have dwelled so far.

The first cornerstone is the ontological proof, the Anselmian argument that, with a stringency never accomplished elsewhere, has connected in an indissoluble link thought and thought of God, thought of the absolute and existence in thought of such an essence. Anselm tells us that there is a place, and the place is God, where essence and existence coincide or, to say it with the terminology we have adopted, where existence has *always already been there*, inscribed in God's very essence. This is the starting point, the ontological starting point. Yet how should one handle such evidence when the coincidence occurs in a subject that we are not, and that is even that which we are not, because we are finite and such subject is infinite? How can one subtract it to doubt, which modernity, with Descartes, has entrusted with the foundational task of ascertaining the truth? One can only reverse the journey and start from that which *is always already there* (without being there ever since the beginning, without being able to say such an "ever since the beginning"), that is, from the I and its existence that is already given and cannot be subtracted.

One starts from the I, in the sense that the I is the presupposition that cannot be left out. Descartes has taught this to us with no subsequent real refutation of his approach except in a pejorative sense by Hegel, who asserts that that from which philosophy should start is being (a being in the end interchangeable with nothing). Hegel's direction is an extreme attempt at eliminating all presuppositions from philosophy. Descartes' direction is conversely that of considering the beginning of philosophy as that which always occurs *post festum* [after the beginning]. The attempt at restricting such *post festum* to the minimum that is possible does not have a minimalist meaning because the I (from which one starts) is not an almost nothing, but rather an almost everything. It is that which one cannot set aside (in this sense it is the minimum), but it is also that which tries (without succeeding, as we shall see) to be self-sufficient (and in this sense it is also the maximum). This means that the philosophical exercise begins with an always already there, starts after the beginning, and, despite everything, considers the subject of such a beginning as good material for the continuation of its work.

The I that is thus acquired is finitude. Finitude, which is achieved from out of the I (which would like to disclaim it, and which in any case does not presuppose it), is finitude in the proper sense, because it is not a presupposed structure of existence but rather that to which existence is forced. The finite is such precisely insofar as it cannot define itself by itself; rather, it is interruption that emerges from the encounter with others and with something other, namely, the world. The finite is neither a presupposition nor an existential but rather an outcome. It is an outcome endowed with history and unfolding, and thereby so much harder to grasp and fix. It is not a given but rather something that one finally reaches. It is for this reason that the task of a philosophy that devotes itself to the finite is so arduous; it prolongs itself indefinitely, constantly in a status of search. The overabundance of the finite lies in the fact that, as is the case for tales, any expression "the end" [*la fine*] can be followed by the opening of a new tale and a new variation. The end is reached only at the end. The same is true for the finite, which is an interruption that is reached in the end, a fact of which one finally becomes aware, indeed, an outcome.

From this perspective, the I is to be regarded as a turning point. The I cannot be inhabited; one starts from the I and reaches the I. Within such a double movement, the I designs its own space. It is an intermediate space (which is called "existence" and which is the existence corresponding to the essence of the I, as punctual and precarious as the latter). It is a space modulated by each I in an absolutely original and individual manner. Such an intermediate space is the result of numerous conflicts; it is a diffraction and a dislocation. The I is also the ability to invent within the (mobile yet no less binding) obstacles that constitutes it. The I is not an interchangeable item because it finds and defines itself through the projects, actions, sentiments, and reasons with which it handles the obstacles that it encounters. Not for such a reason, however, is the I immutable, for the obstacles that come across the I have modified and modify it. The I is in its story and, more generally, in the intertwining of the story. Nevertheless, for the I there would be no story if every time again the I were not to refer the contents of the story to itself.

One must start from the I, we said. The post-Cartesian modern has been the grandiose attempt at starting from the I and building worlds on it, [an attempt] not only at building existence, as we have said, but also at modeling history and society after such an existence. What should we do, when [history and society] give us back a broken existence? When the I produces not worlds but horror? Celan's I is Descartes' I after modernity, the I to whom no longer the God of the ontological proof but rather "No one" corresponds ("No one moulds us again out of earth and clay, / no one conjures our dust").[29] It is the I who has lost and is lost, the I who remembers and forgets, the I who is searching for a me, and perhaps even for a thou, the I who is neither self-certainty nor relationship with the other, and yet who wishes to find itself and thereby configures itself as encounter, sought-for encounter, possible encounter. As Celan shows, in the end, one returns to the I even after the horror, because there is no other available starting point from which to begin again except than the assumption of our own history. If Descartes can be understood as a form of secularization of Anselm, in the sense that his *cogito* can be considered as a reproposal of the ontological proof *a parte finiti* [from the side of the finite], Celan is in turn a secularization of Descartes (in addition to Anselm, of course) because the I is not that from which one starts, but rather that to which one is forced to return; it is that which, albeit in pieces, alone is left. The possible overturning starts from here.

The decisive line, the one that produces the overturning, is not here though. It is an imaginary and yet very real line, because the overturning attests to it. Benjamin has given it the name "threshold." In our meditation, we have started from that name. In truth, one can never possess the threshold, one cannot detain it, nor can one properly see it; rather, one attests to it and acknowledges it, or better, one recognizes it as something that has always already occurred earlier. The infinite is perceived in the past, as that from which we are touched, and that has thereby revolutionized, one could even say produced, finitude. The moment comes (and when it comes, it always appears as always already occurred, like the true, which is always new, fresh, and unexpected, but that, when it appears, always configures itself as that which had always been there) when reality is read in reversal: need takes up the name and dignity of desire, and on the threshold finitude uncovers itself as infinity.

The structure that renders all this nonabstract and not merely logical-formal is temporality. I mean that only time makes saying such a dialectics (and even before, it renders its happening) possible. Philosophy must therefore nourish itself with temporality and astutely avail itself of it. The more time [philosophy] has, the more opportunities and occasions it has to let finitude unfold, extending its limits and declining itself at their limiting. Philosophy takes time, gives time to time, procrastinates, lingers, unfolds, protects, and spreads the present. Time has the dimension not only of an unfolding but also of a return and a project. Time is attention, memory, and expectation. Time is times. In it, simplicity is always complex. I am immersed in time, yet I am never included in it because I always ecstatically overflow it. Time urges me, yet time is suspended. A thousand years are like a day, a day is like a thousand years. Time is real, objective, and stranger: it passes and I grow older. Time is gift, grace, unrepeatable instant: *kairos.* Yet (sacred and profane) time is always and only insofar as it is my time, the time that an I has appropriated.

In a gesture there are infinite, diverse, and simultaneous movements intersecting one another. In order to say them, philosophy must unfold the time of which we can avail ourselves, that is, the present. This is the time from which we need to start again so as not to lose the past and annihilate the future. Moreover, this is the time that can be up to the temporality of time, the time that can respect the duplicity of time's lingering and fading. Such a time can return to being both instituting and a form of institution. For philosophy, this is the gathering, in the form of sense, of the punctual meanings with which the present is interwoven. As we have tried to show, this indeed happens in tales.

CHAPTER SIX

The Tale of Finitude

THE TALE OF FINITUDE

The coordinates we have followed are those of a nautical map—that is, they are useful for a journey. The indefinite surface of the sea is in fact classified first of all in function of the journey. This does not mean that [the surface] has no relevance or meaning in itself, or that in each of its single parts there is no individuality (for everything, in the universe, there is individuality). Rather, it means that meridians and parallels, astrolabes, and nautical maps are useful not to describe the indefinite immensity of waters but rather to fare upon them.

Likewise for finitude. It is immense and indefinite. From a birdseye perspective, it appears always the same; yet, it is always different and even ungraspable for the one who wishes to relinquish oneself to it. For this reason, too, narrating it is the most opportune path.

Novels, these great anthologies of existence, owe their power not to the mass of stories they narrate but instead to their ability to dilate existence. The tale of the novel is not its story, but rather the incredible dilation of the story. I mean that what is truly extraordinary in narrating is the unhinging of times (an unhinging that saves; therefore, the unhinging is also the reorganization of times). Entire days go by as if they were instants; years pass with a simple change in chapter; conversely, instants are stopped, decomposed, retrieved, and retraced in all their details. Time is saved because it does not pass as indifferent and always the same. Rather, it bears with things. It seems to bend to them rather than grinding them in indifference (a trace of this remains in the time of the holidays,[1] which are a stylized present, the stable suspension of the daily flow, and wrapped around a memory and a dream in order to give them strength and receive from them a spur).

Tales are saving because they release the infinity of the finite and do not abide by chronological temporality. Years carry out a destiny; thus, they unfold and deepen rather than surprise. In this sense, in the ideal story everything is already contained ever since its beginning; yet, everything comes to light little by little, as if in a complex and felicitous interpretation. Not facts but time is the instrument in the hands of such an interpretation, not unlike what happens to the orchestra conductor, who does not change the notes yet is lord of time.

Time is the true content of the tale. The only time we can cross is the time of finitude, a time that indefinitely prolongs itself and yet is definitely circumscribed (after the beginning and before the end): such is indeed the present.

No writers dare situating themselves before and after their tale except when, in a stylistic procedure, they create the background for the tale, which is in turn another tale. In the extreme, writers place themselves outside the tale, and contemplate it from the outside, thereby better emphasizing its intimate and original fate. Everything takes place within given and nonprolongable boundaries; yet, everything is retraced according to a time different from that of facts; therefore, everything is also scanned according to a different, free hierarchy of values.

What follows for philosophy is the suggestion to devote itself first of all to the tale of finitude. What one should avoid in such a move, however, is the presumption of an ontological constitution of the finite in the spirit of classical metaphysics. Analogously, one should shun the opposite danger, that is, a mere description of the finite, an introspective description of its spiritual states, as it has ultimately been done by nineteenth-century spiritualism. In the twentieth century, phenomenology and existentialism have come very close to the orientation I intend to suggest. Yet they have perhaps worked still too much in terms of the pair description-constitution, whereas the path that makes use of the pair narration-interpretation seems to me to be more fruitful.

What in fact inexorably escapes the former [pair] is what is most peculiar to the finite one wishes to encounter—namely, time. Only time as presence is in fact accessible to description; and, as Heidegger's analysis has shown, the only way to escape such unilinearity of time seems to be that of unfolding it as an (albeit novel) ontology of time—that is, devoting oneself to a reconstitution of being as time. Despite the acumen of the analysis, what becomes accessible through this path is a structure of finitude that immediately turns into material for a new structure of being. In this way, the indeterminacy of the finite is entirely lost, and through a process of reduplication in the end the world of finitude appears identical with the world of being.

Narrating the finite means instead following it in its folds while retaining time in one's own hands, letting [time] flow or accumulate according to a choice that is entirely up to the interpretation. The Cartesian I is incapable of interpretation and capable only of certainties, because its geometrical universe fixes it in the form of punctuality. Although it is not simply thought, the Cartesian I lacks time. The true turn consists in making another temporality germinate from the subject; this indeed occurs in the form of the tale.

I have said "another temporality" because we are in front of neither a unique nor an absolute temporality. We are faced rather with a temporality that intertwines with the continuous bass of chronological temporality. The time of the subject is not the time of constitution but rather of interpretation. It does not give rise to facts nor does it make the meaning of that which it encounters emerge. Instead, it inscribes facts and meanings within a project of sense; that is, it reorders time according to freedom, fixes its own present, and pivots around it up to the risk of failure. It protects and saves the instant, condemns what is meaningless and accelerates its consumption, and opens new worlds by hypothesizing connections and sources. It shows that there is not only one, but rather many worlds, and that there is not only one but many times. One does not move from the one to the others except through a swerve of freedom that is due to the I from which, through a swerve, a different time indeed originates.

The theory of eons, which Pareyson enucleated thereby reproposing an ancient and unusual doctrine, has the merit of pointing out how there are many orders of reality and how thinking continuity and discontinuity jointly is possible.[2] Continuity presides over succession within the same order. But this order is not the only one. Between one order and another there are instead gap and discontinuity. Time contains the order of succession and the order of disconnection. Time flows, inexorable and anonymous. Time can be halted and anticipated. Memory and desire save the time that is no longer and the time that is not yet. The I is the intersection within which the order of succession and the order of disconnection cross each other and collide. The project of finite existence arises out of such collision. The organic and historical time of development and the contracted time of the project delineate the ever precarious order within which we exist.

There are moments within existence when one time prevails over the other. Like an animal, one quietly accumulates succession so as to have material for one's project, and then one squanders in projects that which one had accumulated, until one learns to live with both—deceiving mere succession by channeling it into a project, and having the project collide with a possible succession. Existence writes itself in this manner, like a book that proceeds one page after the other, toward unexpected and at times even unhoped-for turns. This means that existence is no more project than accumulation, no more activity than passivity. Existence is first of all existence, that is, articulation that within time nourishes itself of accumulation and makes itself a project. Yet neither of these defines existence because, on the contrary, it is existence that defines the temporal nature of project and accumulation.

The narration of existence is done the same way in which Shahrazad narrates imaginary tales, one night after the other—the only purpose being that of seeing a new dawn. One narrates so as to continue the order of succession. This goal (the goal for which the I never interrupts itself, always wishes first of all itself, and wants to live) is, however, not the aim [*il fine*]. Narrating, something happens. The I halts or displaces itself in a different manner. Existence as protention opens a space and situates itself in it—another world, a different time, a new eon.

The tale is long, and can be interwoven following numerous threads. In *Nonostante il soggetto* (Despite the Subject),[3] I have retraced the thread of desire and need, patience and attention. It is the thread of Narcissus who follows Echo while in love with his own words, which are returned to him in a truncated form. Thus he continues to follow himself up until the catastrophe. It is a thread that is exposed to becoming lost but also to finding itself while looking for itself. It is a thread in which the impossible possibility that by looking for oneself one may find another is observed in many of its variants. The decisive turn is when, by exercising freedom of interpretation, time gets to be halted, interrupted, and a pause and a disjointure are created. In it—in front of it—the possibility of losing or finding is given. The face-to-face that one has always avoided could be there. Should one continue to avoid it, or should one accept the gap, the opening of another time, which is not the time of succession alone? There is no cogency whatsoever in such a move. The move may proceed within continuity, like Narcissus who sinks into the waters, loses Echo, and loses himself, or it may assume a new position and give Echo a new name, in which the coming back of words contains also a trace of the origin. There is more than one possibility and, like Chinese boxes, such possibilities open up new games and new scenarios. Time prolongs itself, chances are multiplied. As if in a game, the win turns into a new bet. Finite existence becomes multifaceted and multiplies, enriching itself. Only at the end, when the game is interrupted, will one be able to know if the stakes that have been won are useful for something other than simply continuing the game, if by lingering one has won not only the bet but also the game. In any case, one has not lost, because the game that has been prolonged is in all cases life, and it is worth being lived.

There are still other possible paths of narration. The I that in responsibility devotes itself to that which is needed is in the arms of the other ever since the beginning. It need not look for the other, because it has already encountered such other ever since the beginning. The I has displaced itself there, in exteriority. Here, too, the long path of dedication and commitment prolongs finite existence into austere meanings that are full of dignity. The decisive moment remains, however, the face-to-face with oneself at the end of the day, when the door of the others is closed. If it finds itself, the I does so not in the succession of continuity because, were this case, the I would only find self-reproposing alterity: querulous, interfering innocence. Conversely, if what presents itself once the doors of the alterity of the others is closed is alterity to oneself—that is, alterity to one's own nonbelonging and estrangement—then the I belongs no longer to the order of responsibility because the I stands a step before [responsibility], stands there where [responsibility] nevertheless springs and originates. The I generates worlds it does not inhabit, lives a succession that is its own and yet does not belong to the I. Is such a gap salvation or destruction? Does it annihilate or does it, on the contrary, open the order of responsibility? If the I unfolds the alterity that the I is to itself into an alterity to which it devotes itself, that is, into an alterity where the other has a consistency that cannot be impacted by the I and to which the I turns; if the I opens the order of responsibility as a destination, does the I find or lose itself? Does it accredit and justify the ethical commitment or does it simply explain its genesis psychologically?

Like all differences, the gap escapes grasping, and decisions have no previous justification; they only receive a risky confirmation in the very act of their realization.

The same is the case, infinitely, for the mediation and measure that rule commitment in the world of the "it"; or for the inebriation that enraptures in beauty; or for one's self-entrusting to the merciful alterity of God who comes toward us. In all these paths, worlds open up that are accessible only to the I, but in which the I always places itself outside itself. In political mediation, the I agrees to appropriate and give measure to that which comes from the outside. This is its dedication, but it is a dedication in which the immense ego in which the I assimilates the world does not belong to the I, and in which success itself generates perdition. The ego is not the I, it is not its [I], and it can even be stranger to everyone, like a monument without a figure, a project without subjects. In aesthetic rapture, the I, to which everything nevertheless refers, has no consistency outside its form. The work wins over the subject because the work is the subject that has truly succeeded; however, it is thereby also the failure of the subject that as such does not succeed in being work. Finally, in religious reliance, could the God-for-us, who mercifully bends down and welcomes the I, not be the projection of a heart that wishes to be greater than itself and, failing to be so, passes itself for a welcoming alterity? But precisely thereby, having dispossessed God of God's most precious being, that is, his being greater than itself, would [such a heart] not immobilize God in an immense and uncontroversial presence?

These are all failures, or opportunities of a second return to the I. They are all modes of experiencing a gap, a plurality of worlds, through which one can pass or that one can choose. The world can only be gazed upon from one world alone; the phantasmagoria of many simultaneous worlds is the failure of the kaleidoscope, which fragments and gives rise to simply random forms. Once one has inhabited different worlds, one must choose one of them as the place from where to look at the universe, as place of return and resistance, as fabric for interpretation, as viewpoint of the tale. There one holds on to the crossing of times and worlds; there one risks the beginning of the tale; and from there one measures the width and depth of the arch that one wishes to stretch.

The world from which one starts such narrating is not the world from which one comes in the sense that its parabola has been modified through its intersecting other parabolas. The return is not retrocession to what has already happened but rather anchorage to what has been. Being is always said in the past tense, yet it is never gained by means of a retrocession. From that point one does not pass to an elsewhere; rather, one becomes capable of thinking the elsewhere. This is not only the in-another-place [*in-altro-luogo*], but rather the absolute and proximate elsewhere; not the hard and yet infinitely revocable limit, but instead the flexible yet irrevocable border, which restores the I to the I, puts an end to tales as simply entertainment and evasion, and starts the tale that secures time so as to have time and from there let the time for new tales originate.

FINITE EXISTENCE

"During this period Shahrazad had had three sons by the king and when she finished the story . . . she got to her feet before kissing the ground in front of the king. 'King of the age and unique ruler of this time,' she said, 'I am your servant and for a thousand and one nights I have been telling you stories of past generations and moral tales of our predecessors. May I hope to ask you to grant me a request?'"[4]

We are at the end. Night after night the king, hurt by his wife's betrayal, had taken all the maidens of his country and decapitated them, just like his wife. Only Shahrazad had been able to stand up to him for a thousand and one nights, thanks to the charm of her tales and the cunning of the one who had approached the king of her own will. Every day there was a new story to calm, during the wake, a desire that was not yet fulfilled. This occurred a thousand and one times—in other words, to infinity and something more. Shahrazad's shrewd wisdom has reached the measure prolonging the threads of her finite existence through the delaying gestures of the tale.

As anyone else, she has done so under the always ambushing threat of "the destroyer of delight and the parter of companions."[5] In the face of death, existence encounters its own finitude. Yet it encounters it not as its own limit, as a limit of existence, but rather as an imperious barrier that comes from the outside. It is not something that pushes [existence] back into indigence and need; rather, it is something that inventively exalts [existence's] own power and fullness. The self-announcing death restores one to life as to the most precious gift. Caducity, which gets to be discovered as a barrier one is not allowed to overcome, restores one to the value of what is, rather than condemning [what is] to disvalue. Too long has the finitude of living been understood as a woodworm threatening the consistency of everything that is encountered, failing to realize that what is at work in the woodworm is not the awareness of the finitude of the finite but rather a bad absolute that, in order to assert itself, needs to annihilate all autonomous reality and consistency. The finite is that which by itself is full and fades away once it has reached its own measure.

It is not by chance that tales have multiplied within such a framework and have channeled even different traditions and people. The certain sign of death multiplies the energies that prolong life. As in many other cases [beside Shahrazad] (for example, in Boccaccio's *Decameron*, which was written under the sign of the plague), this arouses the need for narration. The natural frame of death gives tales their authoritativeness, as Benjamin has noted: "Death is the sanction for everything that the storyteller can tell. He has borrowed his authority from death."[6] In sum, the tale speaks of a transient, yet not ephemeral life.

The accentuation of death should not deceive. Albeit marked by mortality, the life that the encounter with death gives back is not defined by death. On the contrary, it is defined precisely by its struggle against death, a struggle that delays [death's] inevitable approaching through the fullness of the accomplishments. That night, when the king granted his promise and, rather than desperately looking for an absolute revenge, gave himself over to the flow of days, thus accepting mortality—that was "a night that stood outside the ordinary span of life, whiter than the face of day."[7]

The Biblical tradition, too, when it speaks about human life, does so while exalting its simple fullness: a bride, a vineyard, an innumerable progeny, a circle of friends. These are things that are always ready to hand and make a life full—a life that can fill up its measure precisely because of its finitude.

The first conversion, which is the most difficult of all because it does not allow any escape into the ungraspability of an absolute horizon, lies in bending oneself toward the finite (our finitude) and slaking one's thirst in its fullness. Finite existence is existence within time. It is not by chance that in the tale the king is said to be the "ruler of . . . time."[8] Prolonging Shahrazad's nights by granting her a new adjournment is up to him. Although he has the power of prolonging or interrupting, the ruler of time nevertheless does not himself have the power to give joy to his own days. He must obtain such joy, placating his senses, from the young maidens whom he will then kill because in the brief turn of a night he has emptied them of all their worth. Only the instant is precious—the instant that is short-lived, unstoppable, lost. This is not the case with Shahrazad, who prolongs instants into the tale and generates him children. With her, finite existence begins. It is not the absoluteness of hatred, nor the despair of the instant, but rather the rhythmic flowing of time—the present, at last.

The finite that is, escapes; the finite that escapes, is. Like in the contrasting rhythm of respiration, caducity tightly interweaves accrual and dispersion, and gives birth to a precarious stability that is our whole good. As it has been said, philosophy is born out of wonder, out of the amazement that something is—that being is. It is born, therefore, out of the encounter with the consistent caducity of the finite, because of no other being could one speak in wonder. Neither nothingness nor the being that necessarily is arouses wonder; rather, albeit in different ways, they annihilate and cause horror. On the contrary, regarding the finite being one is surprised that it is, and one is grateful for its being.

Such an origin marks the path of philosophy. Even the dizziest speculations contain the hidden treasure of such wonder as a good to be protected. The love of wisdom by which one allows oneself to be fascinated does not lead elsewhere, far from wonder; rather, it leads by its abode. Love of wisdom does not deny but rather protects the finitude of human existence.

One could say that we do not know such an Edenic finite that disappears only once it reaches its own measure. Rather, we know a finite that is threatened by injustice, besieged by pain; a finite that itself produces injustice and suffering. In a word, we know a finite existence crossed by the great mystery of evil—the evil that it realizes, the evil that it suffers. We know a twisted finite. Only blindness could ignore this. We need to continue thinking of the finite and of evil, however, which we know as factually joined, as distinct as to their principles. The ancient tale of the Garden of Eden has been used too often to solve the problem of the imputation of evil, and to answer the question of the origin of evil in the world with one name; instead, it should be rather thought in terms of its reserve of goodness. It testifies an intact finitude that is the origin that is more originary than all finite. After the fall—because all tales are written after the fall—the finite maintains memory of an intact and finite origin.

In this there is a utopian reserve from which one can draw anew. This is perhaps the homeland of which Bloch speaks when he hints of a homeland in which we never were but that shines in everyone's heart. It is neither the conceited pretension of self-sufficiency, nor the myopic denial of the evidence of evil, nor the cynical marginalization of limit and need, but rather the demanding holding on to a lost yet not forgotten origin. Every time we patiently work at restoring the finite to its dignity we again encounter the origin and testify that the origin, albeit threatened, has not disappeared.

To find it, one needs to dwell by the finite.

All this is in peculiar consonance with a religious perspective. In the Book of Kings one finds the tale of the prophet Elijah who, at God's mountain, Mount Horeb, is addressed by the Lord who orders him to come out of the cave, in which he has sought shelter, and to stand in the presence of the Lord: "And behold, the Lord passed by," says the holy text, "and a great and strong wind rent the mountains, and broke in pieces the rocks before the Lord, but the Lord was not in the wind; and after the wind an earthquake, but the Lord was not in the earthquake; and after the earthquake a fire, but the Lord was not in the fire." In none of these manifestations of power has the presence of the Lord revealed itself. Yet, "after the fire [there was] a still small voice. And when Elijah heard it, he wrapped his face in his mantle and went out and stood at the entrance of the cave" (1 Kings 19:11–13). Even the Lord's presence has more to do with the frail finitude of the light breeze than with the vigorous violence of a forceful wind.

INFINITE EXISTENCE

The finite is an unstable object. It is so unstable that it escapes from all sides. When a human being encounters death and discovers that existence is finite, the long, timeless summers of one's childhood too come to an end. Only some moments of solitude—the torrid urban heat outside, or the endless extension of a suddenly unmoving nature—bring back, unexpectedly, that lost feeling. For the most part, instead, the flowing of days undergoes an upsurge and as it were an acceleration, and finitude sinks into an endless replenishment. The precariousness of the finite almost inevitably leads it to an infinite fulfillment. Dwelling within the limits of finitude without becoming either afraid because of its fugacity or proud over the indefinite ability to elude one's own identity is hard.

The finite escapes. It is the most difficult object for philosophy, because when, as reflection, philosophy goes back onto its object, it finds it to be no longer the same. The finite is constantly traversed by something that we can *provisionally* call "infinite"—the infinite power of feelings, the infinite overabundance of nature, the infinite abyss of death, the infinite passing of entertainment. Around finite existence, everything speaks of the infinite, which is the most naturally first object of perception.

There is no reason to rejoice at this, though, as if human beings possessed a privileged access to the infinite. The infinite that meets them is nothing else than a distorted image of themselves. Narcissus and Echo are figures of the same tale. Echo is not allowed to disclose her own passion; in love with Narcissus, she follows him around without ever being able to address him with her own words. She is allowed "only the power out of many words to repeat the last she heard,"[9] so that there is no communication whatsoever between the two, but rather a reciprocal appeal that does not become conversation. "Is anyone here?" Narcissus, lost in the forest, asks. "Here," Echo equivocally answers. Without a name, it is an answer that sends back the same question. "Amazed [Narcissus] looks around in all directions and with loud voice cries: 'Come!'; and 'Come!' she calls him calling." [10] Whereas Echo, who would like to speak, has no words of her own, Narcissus, who would like to reach the image he is looking for, finds nothing except his own image. The face he encounters on the uncertain surface of the water is nothing more than his own reflected and repeated image, as reflected and repeated is the word that Echo utters.

The dream of the beloved other inexorably returns the face and words of the lover; the beloved hides or better gives herself in the form of a duplicated identity. In the same manner, the infinite that the finite encounters is nothing else than the finite itself, duplicated and dilated to infinity. In this sense, Feuerbach's criticism is correct in its showing how humankind has projected its own essence into the concept of God and has hypostatized it into a being that lives of its own life. What is lost in this way is not the infinite, however, which in itself does not appear to be denied but only diminished by this procedure; what is lost is the finite. Dilated to its own indefinite power of production of the infinite, the finite is no longer what it is, because it is cast into a race the law of which forces [the finite] to be constantly outside of itself, projected into an indefinite self-overcoming. It is evident that on this path the finite will find the infinite only as a projection of the finite; at the same time, though, the finite will also fail itself, because it can meet itself only by never finding itself, like the word that is given back by the echo or the image reflected by the mutable mirror of water. Constantly stretching in the direction of the infinite, the finite does not encounter but rather loses itself (that is, it also loses itself as finite).

Existence is instead traversed by the infinite in a different manner—not as a projection but as a wound, not as a fulfillment but as an interruption. The infinite is that which the finite is not, or better, that with which the finite is engaged in a never-ending hand-to-hand struggle. The Bible preserves a trace of this in a mythical form in a tale that the exegetes unsuccessfully wish to attribute to idolatrous residues; in fact, it is the untimely fragment of the presence of the infinite to the finite. It is an enigmatic and mysterious tale that, in the desert night, opposes a Stranger to Jacob. There, in the land of Canaan, just beyond the Iabbok, Jacob, who is now alone, is confronted by a man. Violating Jacob's solitude, such a man starts a fight that is destined never to end. In fact, it will have neither winners nor losers but only questions. Meeting and fighting, the two have exited their solitude. "What is your name?" the Stranger asks Jacob. "Tell me, I pray, your name," Jacob answers. At the appearing of dawn, that place has a new name; it is called Peniel, because there Jacob has seen God face-to-face and has had his life safe. When Jacob was passed Peniel, the sun rose. He had fought with God and had not lost; only now, he was limping because of his thigh. That God who had lamed him, though, had also blessed him and given him a new name: Israel—for he had been strong against God and the human beings.

This page (Gen. 32:23–32), which has been haunting me for twenty years, has secretly guided my research. For the encounter with the infinite being or, to use the religious name, with God, is first of all the encounter with a stranger who blocks the way, with a "no" that interrupts the indefinite self-dilating of existence, with an obstacle that one rightly contrasts but cannot overcome. The infinite is not the strengthening of finite existence; it is what blocks it—an insurmountable barrier.

Under the mark of such a great negation, however, the finite is restituted to itself in its own finitude. The encounter with the infinite, which is conflict, allows the finite to recognize itself in its own precarious yet not useless balance. It is not necessary to emphasize the power of dilation of the finite, as if [the finite] were not eventually to encounter an obstacle that it cannot overcome. Nor is it necessary to mark the finite with the mark of a limit, which in truth, in its own spontaneous stretching out, [the finite] does not know except for when something from the outside stops its own further stretching out.

This way the finite is returned to the measure that defines it: in its dreams, in its projects, in its works, [the finite] is always more than itself, always less than itself; it is always elsewhere than where it is, and never in such elsewhere; it is always weak and never defeated; it is always strong and never full.

Infinite existence intertwines with, rather than fulfills, finite existence. It interrupts yet it does not block it. It defines without constituting it. Thus, the finite finds itself. The theme from which we started ceases being a need and becomes a possible reality. It is true: Without the infinite the finite would not be, but it is not for this reason that one should imagine an annihilation of the autonomous consistency of the finite. On the contrary, precisely in such consistency, understood neither as the explicit fulfillment of the implicit nor as its denial, it is possible to retrace the nexus binding the finite to the infinite. It is a nexus of reciprocal difference up to the point of conflict; however, for the finite, it is also the nexus of its own belonging to the nexus of difference. It is the path to reach one's own identity while subtracting oneself from mere self-projection or duplication of oneself.

PRAISE OF LINGERING

For the finite, the matter is that of holding on to its own consistency, on to that being that the finite being is. This is what I name "resistance"; in the German term (*Widerstand*), there is at the same time the tension of the conflict (*wider*) and the pride of the one who stands (*Stand*) and holds. Resistance is not, however, a frontal fight in an open field; it is rather a fatiguing and daily laboring. In front of itself it has not an enemy with whom to engage the decisive battle, but rather a polymorphous adversary: time. Resistance is the struggle of holding on to that which has value, of building a sense on that which caducity threatens with disappearance. It is therefore constantly exposed to a conflict with time. There is no way to defeat time, though, except than through a stratagem: by favoring it.

The following may sound like a wisdom-oriented anecdote, yet it is history of philosophy: to the one who wishes to escape the bind of noncontradiction, Aristotle opposes no argument except the invitation to say something, anything.[11] Once the opponent is chased out of silence, which assimilates him to a stolid plant, he is already defeated because anything he says, he has already submitted to saying something and implicitly to denying that which contradicts the former. Something similar occurs here in the case of time. Time and the immense/weak power of its flowing have always already won against those who claim to escape them in favor of a time outside time, or to oppose them in favor of the time of eternity. If it is not possible to contrast directly the principle of noncontradiction, against its bans it is, however, indeed possible to exercise the art of contradiction, as the entire history of thought shows in its being committed, within the horizon of an undeniable identity, to saying something that is not reduced [to identity] and that therefore forces [such identity] beyond itself, toward alterity. The same is true for time, the mastery of which can be suspended precisely by favoring its unfolding. Giving time to time, offering a chance to the present are the ways of conquering time. Time that has unfolded, that has been allowed to develop, is time that retains its caducity, if it does not contradict it. It is time that gives fragments the opportunity to display their value. In this manner, resistance takes up the form of lingering.

Unfolding, time stays by the finite and does not annihilate it. Moreover, when staying in this manner, time allows the finite to get shaped in the form that belongs to it, that is, adding one term to the other, summing a finite measure with another finite measure. The multiplication of single points is not in itself the acquisition of a horizon. The accumulation of meanings does not make up sense. Nor does the sum of remembrances produce memory. [Remembrances] could even impede [memory], in the same way that the proliferation of meanings may generate the elusion of sense. Yet, they are each the condition [of memory and sense respectively]. Lingering means then enabling the conditions that let the finite reach itself. Doing more is not possible; doing less is not licit. As in the evangelical parable, the anticipation of judgment is a temptation: separating the wheat from the chaff is neither licit nor possible yet.

In lingering, the great possibility of the penultimate, of that world for which we are all called to work, unfolds. For this reason, I think that philosophy must return to meditate on the great motive that marked its origin, namely, the love of knowledge, and it must be the concrete exercise of it. In doing so, [philosophy] rejects the temptations that accentuate its status as science, but also those that confine it to being ideology of private feelings. Love of wisdom is a formula dense with tension, because it binds philosophy to wisdom, and hence to truth; yet, it forbids possession of [wisdom] by prescribing love of wisdom, perennial search for it.

Is there not, in this precarious definition, an unsustainable contradiction, which is inevitably destined to be solved into the primacy of either a self-fulfilling search or a knowledge that puts an end to the search? The answer is "yes" if philosophy constitutes itself under the sign of anticipation, that is, as culture of fulfillment. From this perspective, which is the perspective of an unfolded truth and an accomplished totality, the search can only appear as superfluous. Likewise but inversely, from the perspective of the search as self-affirmation of the searching subject, reference to the truth can only appear as a hindrance. Yet truth and search, love and possession are not a hendiadys but a nexus—that is, something that can always and at any moment be separated and parted up to the point of an alternative. From this perspective, they are something frail. And yet they are something that lets their most intimate essence transpire precisely in this structural defenselessness. Although it does not produce the negation of their nexus (which at least in the form of an ineludible polarity remains on the background), the splitting nevertheless produces the distortion of precisely that which is chosen. Search without truth in fact appears to be not unlike truth without search—both are nothing except claims, which vary only because of the intensity of the psychological intentionality with which they are asserted. What differ are only the human types asserting them, and not the contents characterizing them.

On the contrary, precisely the mutual self-entrusting of search and truth seems to allow a nonbanal understanding of what each is. Through an indirect way, as always happens in the case of that which is important and valuable, truth lets us understand something about the search, and the search lets us understand something about truth. What comes to speech, in fact, is that the certification of either can be neither direct nor immediate. Truth does not bear witness to itself; rather, someone bears witness to the truth. Likewise, good faith is not enough to justify the dignity of the search; the "where" toward which [the search] is directed is necessary. In this sense, a feature joining search and truth, love and possession emerges little by little—namely, the fact that they both look for their own guarantee outside themselves. What happens for search and love may seem obvious (such is the force of the movement that animates them); less granted seems to be the case for truth and possession. Nevertheless, on careful consideration, truth as well as possession is defined by that *of which* they are truth and possession. That is, they stretch beyond themselves into the formula that says the truth or into that which reveals possession. In such a way, they seem to declare the search over and the goal achieved. Yet the goal that is still subject to being formulated and the possession that is subject to being exhibited are an enunciation of the fact that both can be always achieved anew; both must be always conquered anew. In order to attest itself truth must, through a potential demonstration, indicate the path that has rendered it possible; possession must exhibit proofs of the labor of conquest that makes it legitimate.

In this form, the nexus is, however, simply a sort of curse, a trap, the impossibility of completely shaking away that from which one wishes to be parted. What keeps together truth and search is a bond not unlike the one each thing has with its own shadow—inexorable and useless. In this form, the nexus seems to be blocked, unproductive, not unlike a heavy burden. One should instead make this link work positively, and show how it does not disclose a bond but rather a possibility—namely, possession measured by love or truth animated by the search. Possession arousing new love and truth generating new search are not less but rather more than truth or possession that are absolute.

To be truly such, [possession and truth] must stop being defined through atemporal immediacy, because only in the full assumption of temporality can the nexus find fruitful articulation. Here is a last reason in favor of the temporality of lingering. Instead of thinking—or despairing—that time is fulfilled here and now; that is, instead of producing artificially the realization of the *kairos*, lingering actively sets itself out to listening. It operates so that the "here and now" may be many, so that the *kairos* (the opportune time), which it is not its task either to produce or to sanction, may be. It does so in the only way that is allowed to it—namely, by accumulating time that is worth being, turning the present into a threshold and not a point, a horizon and not an instant, a sense and not a meaning. Giving time to time by lingering at the side of finitude is not only, reductively, delaying its disappearing; it is the passionate resistance for and attention to the finite [that are] capable of letting the rich, multiple, multifarious dimensions of the finite be.

CHAPTER SEVEN

The Great Tale of Time

THE TALE AS DICTION OF THE PRESENT

The tale is articulation (extension, dilation, restriction, essentialization) of the present in the structural condition of temporality, that is, as something that neither is presence nor is available in the form of immediacy. The material of the tale is actually not the facts themselves (which not by chance have become vicissitudes) but rather time. The articulation of the present that has been mentioned is mastery over time. The narrator is impotent in front of the misfortunes of the world; the heroes are subjected to the freezing wind of occurrences, they may be subdued by them, and they may even die. What is left to the narrator is the speech that unfolds the tale, the announcement, the acceleration, the restriction, and the dilation of the instants. Once it is actualized in the present of the tale, the unfolding vicissitude appears to be saved, if not in terms of its single occurrences then at least in terms of the time converging into the tale. The present that is thus constituted is not only the present of narrator and reader; it is also the present of characters and tale. The time that is saved is likewise and at the same time salvation of that which is deposed within such time. The narrators, who are masters of time, cast the net of time and, despite its wide meshes, which cannot retain much, they nevertheless save something; the narrators save it without changing its genus or its essence or its temporal index.

The tale is the present and not simply the current times [*l'attualità*]. Because it is time, the tale originates from a subtle yet ineludible disjuncture. The tale opens the present as threshold. This opening is itself made possible by the finitude of the present. In the act of its happening, the present is already beyond and after itself. Its extreme finitude and its structural caducity let the alterity of the passing time shine through in the identity of the instant. One can then celebrate the vanity of it all and abandon oneself to the passing of time. Yet the Heracliteanism of melancholy does not hold, because it incongruously presupposes the strong identity and immutability of the one who chooses and proclaims the passing of all things. Alternatively, one can transform the present into presence. One then loses both temporality and finitude. The present that has been rigidified into presence assumes blocked temporality as if it were authentic temporality, and duplicates transient finitude into a structure of being. On the contrary, the tale works on the disjuncture of the present and, in order not to lose it, tries to open for it a (temporal and finite, provisional and imperfect) passage toward permanence to the point of taking upon itself even the chance of always arriving too late, when time is already fading away and can be redeemed only through memory.

The modality to save the transient immediacy of the present is that of unfolding its sense narratively, as something that, having taken up value, is attained and somehow fixed forever though not thereby eternalized; but also as something that has always already been lost because the tale narrates the present, yet it does so when the present no longer has the feature of the current time [*attualità*].

The tale originates from the swerve of both an absent presence and declining present. The tale knows and perpetuates the immanent wound of the time that passes. It bears witness, therefore, to the present and not to presence, because presence that is no longer is the pure emptiness of absence, whereas the present that is no longer current [*attuale*] can nevertheless be acquired as present via the threshold of the tale (in its passing, [the present] confirms only its essential temporality).

To tell the truth, there is also a tale that is born out of superabundance; it originates in fullness, and almost out of an inebriation of presence. In it, a community not so much represents itself as much as it invents itself. This tale, too, originates from a swerve. Albeit in an upside-down form, it also appears as marked by absence: it projects a sense that is experienced without having been understood. In any event, it too desubstantializes and destructures presence because it rereads [presence] in the form of the present—that is, of a present that is not entirely deciphered not even when one perceives, as it were, a thrill of its unfathomable depth. When in the current times [*attualità*] one perceives the seriousness of a moment that is decisive for personal or collective history, this is as if a presage of the present. It has all the features of the present, because it is simultaneously presage of something that remains, but also of something that passes. "O dolente per sempre colui/ che da lunge, dal labro d'altrui,/ come un uomo straniero, le udrà!/ che a' suoi figli narrandole un giorno/ dovrà dir sospirando: io non c'era:/ che la santa vittrice bandiera/ salutata quel dì non avrà" (Forever sorrowful will be the one/who from afar, from someone else's lips,/ as a foreigner will hear (such words)!/ who, when some day narrating this to his children,/will have to say with a sigh: 'I was not there'/ who the holy winning flag,/will not have saluted on that day).[1] (Where one should notice that the ***kairos*** that has been missed becomes curse that is fixed forever because the narration and handing down to the children will have to include also such an unredeemed guilt.)

The passing moment is like the passing Lord, like the wing-flapping of the angel of history, who, as in the case of Elijah's vocation, reveals itself in the light wind. There, while passing, the true reveals itself. There, as Benjamin would say, a *Dialektik im Stillstand* occurs. There the present becomes apocalyptic. Now it remains veiled, however; it is recognized by someone who, like the Lord's disciples, properly will never be able to bear witness to what he or she has seen. This is because what one bears witness to is what the others at that point will have also seen and recognized or what, conversely, everyone still resists and is unable to see. Birth, however, cannot be witnessed to anyone, because we should be capable of rendering one a participant, in equal measure and simultaneously, in both an absolute certainty (now!) and an absolute frailty (is now truly the now?). One of the most intimate meanings of the holiday, which interrupts the flowing of days, is its holding fast on to such an almost incommunicable experience, and its insistence on building a social celebration on it—that is, on its incommunicability! From here, from such a location suspended on a dangerous divide, there derive also the ever-present immanent dangers of a rhetoric of celebration that substitutes itself for participated holidays. What are duplicated here are the same problems that we have addressed in the case of the dialectics between the individual and the institution.

Perhaps the tale alone, in an indirect way and through the long spires of an involving narration, can make one partake in such an experience. This is so also because, quite frankly, the tale does not claim to reproduce but simply to narrate [the experience]: as something that was. Aristotle's definition of the existing substance is wise, deep, and temporal—*to ti en einai; quod quid erat esse.* The imperfect tense [*erat*] is the enduring form of that which remains present.

In either case (in the modality of absence as well as in that of superabundance), the tale is not a mode of representation (*Darstellung*) because it does not aim at reproducing presence. Presence can be mimicked and represented. The present, however—that is, the time that is—can only be experienced and lived or prolonged in the form of the tale. The tale resists its being turned into a show. It remains imperfect, incomplete, and tragic up to the point when it attains the spectator's involvement and, in that way, it occurs. All tales tell of something that exists no longer, yet still is. All tales tell of a present, yet when it has been; that is, they tell of a present insofar as it used to be [*era*]. Through the iterative character of the imperfect tense, they try to ensure the consistency of [the present that used to be], though without losing its temporality. The imperfect tells that which used to be; that is, that which used to go on being and in this way still continues being yet as time that passes. In the imperfect tense there vibrates an iteration that is lost as daily repetition and yet at the same time is—forever—preserved in consciousness. The time lost is both lost forever and, in the tale, forever withheld as a present that is attainable to everyone.

THE BEYOND AS THE ALREADY-BEEN, THAT IS: NAIVETÉ IN FRONT OF US

If existence is always already there, and the real meaning of the ontological proof is the recuperation of the identity of essence and existence, then the beyond is the already-been in which we never were. What we are looking for is that which we already have behind us, because it is we as another; in other words, it is existence as coincidence with essence. This coincidence never was because existence, insofar as posited and not positing, finite and not divine existence, is always *ek-sistere*, to be exposed. The laceration of the origin is not, however, without trace. The long path forward, in always renewed projects, is not an escape from the origin but rather the dilation of the threshold as a form of lingering close to that which we are.

One can understand how, in its most ancient form, the tale always begins with "once upon a time." Through the trick of a dislocation back in time, [the tale] tries to situate itself in an intact time, where existence is still in proximity to essence. Even afterwards, [when one is] much farther away from the origin, one tries to deepen the threshold—both the originary threshold, that in which essence turned itself into existence, and the other decisive threshold, that in which existence unites again with essence.

The really great question returns to the beyond that the essence is. We do not have an answer for this question, which approaches the risk of nothingness, except than through our clinging to the *ek-sistere* that is still something. It is a something that searches for the other and the beyond as if it perceived not only less than it could, but also and always less than it should, less than what belongs to it by essence. Certainly this is a hypothesis, certainly it is nothing else than the interpretation of the one who, lingering next to existence, imagines its framework reflectively. Ontology, of which we can no longer speak in the a priori form of an ontology of the origin, can be nothing more, because it can only be a posteriori; that is, [ontology] is, at the heart of finite existence, the reconstruction of the conditions of possibility (which we wish) of such an existence.

Transcendence has been replaced with transcending; the hierarchical and spatial image of earth and heaven has been substituted by the tortuous retracing of time's turnings. Nevertheless, we still continue questioning in the direction of a beyond, which is perceived as the condition of that which is. Even formulating the question and speaking rather than remaining mute are an emergence from being there [*esserci*]. No one can sensibly turn the formulation of the question into the proof for the existence of an answer. Yet even when the conclusion is the meaninglessness of the question itself, the answer that the question solicits still opens a space within which it makes sense to be—the space of discourse, encountering, and philosophy. Discourse, which unfolds within such a space, does not deserve a conclusion in terms of inanity because such an answer (and even such a sentence) is the only one inadmissible, for it is the one that would declare closed the space that instead has been opened. All this is not without analogy with what has been seen with respect to the principle of noncontradiction. It confirms the admirable symmetry holding together not only the *cogito* argument and the ontological argument, but also this minimal opening of truth and that maximum logical principle. In any event, in the space that has thus opened, answers confront and measure one another: not so much for their ability to answer the question, and hence to close the space of discourse, but rather for their power to enlighten it.

What remains to be asked is whether this circular path, in which the beyond I am searching for goes backward toward the question that ever since the beginning has posed itself as a questioning, is anything more than the self-confirmation of an illusion. Here, too, an answer can be neither clear-cut nor definitive. And yet, precisely because the beyond is behind us, the question does not have the victorious feature of an anticipation but rather the painful character of that which imposes itself on me. Ultimately, I wish I would not question. Life would flow more smoothly without questions. Yet the question imposes itself on me and cannot be avoided in its entirety.

The choice here is either denying and eluding [the question] or consciously and freely turning it into the origin. When the latter happens, however (and this is neither certain nor granted), this can be claimed as the first, precarious and yet stable coinciding of essence and existence, because the existence that questions duplicates its own being-there [*esserci*] into a question, and with respect to itself answers that existence not only is, but also exists. The very nature [of the existence that questions] is present in the answer to the question it itself has posed. Such nature is beyond existence, it never coincides with it. Since such nature is the answer, it always precedes the question, because it is the beyond that is always already before the question. The beyond is an identity in which we never were, but which remains our origin.

If the beyond is behind us, and if as Kraus says the origin is the goal, then naiveté, rather than being that which we are no longer, is that which is in front of us. Schiller is the first to develop the awareness that the naive is intertwined with non-being, and he speaks of it as the sentiment that puts us in contact with lost nature. As is well known, the consequence of this is that the only possible historical condition is that of sentimentality. Yet, still as a consequence, if the sentimental is the historical condition of modernity, in order for [modernity] to have a development it is necessary to imagine, forward and as a propulsive force, the ideal, that is, sort of a projection of naiveté. In this way, naiveté ultimately comes to coincide with originary nature, and narration [coincides] with one's own situating oneself within it.

The tale, on the contrary, has nothing to do with the originary because in this case it would be nothing else than metaphysics of being. The tale is the lingering of finitude. The tale originates from one's no longer being naive, and makes its way toward naiveté; yet such naiveté is not lost naiveté, that is, originary nature, but rather the naiveté that lies in front of us—in other words, ready to hand, the symbolic projection of what we would like to be and could be.

When it says "once upon a time," the tale says the *eschaton*; it never says what is historical but rather what is permanent and therefore definitive. The tale looks forward; it liberates the future. It can do so because it has retrieved the relation with the past. In the tale is the stretching of the threshold that brings back to itself the opposites of past and future, and that imagines a possible naiveté, an available nature, a homeland in which we might dwell. The tale always has a moral of which it would like to persuade us because, since it belongs to finitude, it can imagine only that which is in some way also real. Its dreams are available. Naiveté, which is not nature, is also available. It is in front of us as the nonreached reachable, as the utopia that could be.

The tale unfolds within this dimension: from lost nature to ingenuity, which remains the extreme—in front of us—of finitude. For this reason, the future, or the in front of us, will always have the human, all too human, features of finitude; it could not be otherwise. The human being who is entirely immersed in the finite is projected toward the future; the mundane horizon of history, its omni-comprehensive horizon pivots around the anticipation of the future. Yet when, as in successful tales, the future is brought back to the threshold, the future is brought back to the present, and naiveté is brought back to the sentimental condition; when concentration makes the future converge backward, then the two dispersed fragments—lost nature and naiveté—can encounter each other again, at least for a moment. The beyond is no longer sought forward, but rather folded backward. The project understands its own being memory, and is no longer satisfied with being naiveté. The eschatological "once upon a time" folds into a once upon a time when we were not but from which nevertheless we come.

Out of this encounter and at the same time, the beyond proper to the past acquires a new and more correct character: it is not historical regression toward that which has already happened but rather recuperation of the ground. The only affinity existing between historical past and past of the origin is that the consistency of the origin rests on being. Since such being is that in which we no longer are, the being of the origin is its having-been. The having-been is, however, not the having-occurred. Although it shares with the past the modality of a consistency that may be eroded, the having-been belongs to a different order than that of the having-occurred. The naiveté that within the historical order is sought forward, across the future, refers backward toward the beyond that naiveté itself is no longer; that is, toward lost nature, which has not happened. Therefore, it cannot be narrated. The tale stretches itself from the historical future to the historical past, yet [it does so] in the sense that it brings them back to the threshold of the present. It brings the naiveté that has become ideal back to nature. One cannot expect, however, that the tale, not even the tale of philosophy, may reintroduce one back into lost nature, into the origin. The tale fills history up to an unlikely point; it dilates it until history in its whole, and not only the present, configures itself as threshold. The tale travels across history far and wide; it dilates and concentrates it, gives it value and makes it explode; it orders and disarranges it. Time becomes a threshold. At that point, the passage to the origin becomes possible. But we are already beyond the threshold. Regarding the passage, then, there is no tale. It is the end of the tale, the last word, which, after one thousand and one nights, restores one to life, to the always lost but never forgotten nature, in whose name we consider our staying not as a dwelling but as a standing on the threshold.

Philosophy remains a penultimate word because it does not exceed the threshold. Still, to neglect what it sees, hears, and comes back to tell us from its standing on the threshold, from such an extreme point, would not be generous. It would not be generous on the side of the one who does not dare to get close to such a border, and remains prisoner of a filling but unsatisfying factuality. Nor would it be generous on the side of the one who, no matter how one arrived there, considers oneself to be well beyond the threshold. In order to communicate one's own experiences, the latter is also called to return within such a space, to situate oneself on this side of the essence in which one was, on this side of the naiveté one has attained. Returning, such a person dilates the space of the threshold and conquers new essence for existence.

Philosophy remains penultimate; yet it is and remains close (the closest perhaps) to the ultimate. This very distinguishing is nothing except a mark of the condition of being penultimate, something for which we await redemption.

BEFORE THE END AND AFTER THE BEGINNING: PHILOSOPHY

When it comes to thought, we are never at the beginning. Philosophy always begins afterward. Thus, after having touched various times in this book on the question of the origin, as if to gain momentum for an explicit thematization of it, it seems instructive and not without irony that it reappears now, at the end, in an epilogue.

What modern philosophy has variously perceived—that is, that the origin is not available to us and therefore we cannot treat it as a principle (and the principle is not available to us and therefore we cannot treat it as a beginning)—continues to appear irrefutable to me. On this ground it is equally impossible to solve the beginning into the end. The final, irrevocable word is never said except that in the mode of the extra-philosophical exit from the scene because death occurs and the word is interrupted. In philosophy, one never truly speaks of the end (as well as of the beginning) except through that trick that makes the end coincide with the origin and allows one to begin anew: nature that *iuxta propria principia* [according to its own principles] is restituted to its eternal and new germination.

Furthermore, philosophy does not begin, does not mark the beginning because it is not thought of separation. By beginning, it would mark a line of separation, it would decide, through a violent power that is not available to it, between being and non-being. Philosophy, which intervenes within a history that has already begun, is rather at the beginning. It questions the sense of the history that has begun and that it receives. It perceives the difference yet it does not produce it through a decision. It protects, unfolds, and questions it in freedom. In order to do so, [philosophy] has available only the material that it has found, and nothing remains for it [to do] except for focusing on it.

Everything gets to be decided within such a work, which is eminently a work of interpretation. Philosophy can execute [such a work] only by situating the data that are available to itself within a picture; yet it cannot consider such a picture as something that it possesses. [The picture] is always simply gathered or, as I would like to say, found [*trovato*]. The beginning is neither given nor posited but rather found, where "found" means at the same time invented, with a gesture of freedom, and discovered, as an obstacle that one discovers on one's path. When the double meaning of "found" coincides—that is, when what one discovers on one's path (the meanings that atomically come to constitute the picture of the world) corresponds—with the sense that has been found, that is, imagined as the sense of such meanings, and when reciprocally the sense that is proposed through an act of freedom finds a mode of its own manifestation in those meanings, then one has reached the only beginning that is possible philosophically. It is [the beginning] in which two paths encounter each other. The beginning is an encounter, and not a principle [*principio*]. The beginning does not begin then because the beginning is always already beforehand. The beginning is not a principle because neither is it posited nor could it be presupposed.

It is not absolute beginning, it is not absolute principle; rather, it is the beginning of an encounter (between sense and meanings, horizon and givens) that factually is always simply possible and always revocable, but for which thought has always the need. This is the ontological order, from which nothing descends as though from an archetypal principle, but which is rather the newly found condition of possibility to think that which needs to be thought.

From a phenomenological perspective the origin is simply and only found again. The identity of the subject is not possessed but rather conquered through a mirroring that, when it is nothing more than reduplication, is mortal (the experience of the mirror, the waters of the lake, the voice that the echo restitutes). Conversely, [such an identity] is an opening when the return to the self occurs as a holding on to identity as on to the condition that enables faithfulness to the other (in a Ricoeurian mode, *soi-même comme un autre*). The identity attained through mirroring or, to say it in a more abstract and universal format, through reflection is a sign of the origin not because it confronts one with the origin (almost as if the origin were identical to identity), but because it describes the path of disproportion through which one reaches identity. The beginning is that in which we have always forever been, but that we never experience. This happens because we are, forever and structurally, always after the beginning. This condition of disproportion, in which we are always and that marks even the simplest of the relationships, that is, the relationship of identity, is the phenomenological confirmation of the fact that we are always a step after the origin. The step could not even be thought, however, except against the background of a missing origin. This seems to me to be the salient element. The encounter, from which one begins, attests the relationship; yet the relationship, which is indeed experienced in the encounter, is nevertheless experienced in its happening and in its permanent feature of event (hence the sense of eventuality). It is carved in the sign of an absence. Precisely because it happens, the relationship is not.

What one thinks is not the absolute but rather the finite. It is a finite that is sounded until one recognizes it as having begun and as beginning, as frail and powerful; that is, precisely as finite, always after the beginning and before the end. There is no principle of the beginning; there is a metaphysics of the finite. When one becomes fully aware that we are always after the beginning, and that the beginning can never be turned into a principle, then one becomes simultaneously free from absolute metaphysics as well as from the absolute positivism of facts. One knows instead that, being located after the beginning, one is situated on an inclined plane that cannot find authentic comprehension except through one's leaning out, up to the risky enunciation, through a project of sense, of the difference that is at the beginning and bears no name. I must find a name for it by inventing [such name], so that I can find it as the name that is its own.

The Bible itself seems to suggest something of this sort precisely when it speaks to us of the beginning—and this meaningfully happens repeatedly, various times. Regarding the absolute principle of creation, that is, the origin of heaven and earth, a new principle is proposed to us: the beginning of man and of what surrounds him, to which he gives a name; and then again the beginning of the other human being: of woman, and with her human society. The twenty-six failures that according to the Talmud have preceded creation say nothing regarding the (successful or failed, good or deficient) nature of the created world; yet, they are very instructive with respect to the nature and difficulty of the beginning. In such a beginning, which for us is so important and decisive (which a secular rereading of the world has wished to overshadow, as if the history of human beings were to begin only after the fall), there is a delicate and deep feature that deserves meditation. Eve is encountered by Adam after waking from the sleep. He encounters her as the one who was already there. She is not brought to him in the middle of the day, like the animals. Eve finds herself with the man in the intimacy of awakening, after having been born one from the other in the night that they spent together. At such a time, which summarizes their lying together until a shared awakening, Eve is recognized as flesh out of Adam's flesh, and she receives a name. With delicacy, the sacred text does not speak of an absolute beginning; rather, it begins after the beginning, from recognition and naming. Yet they both have had the shared time of the night.

Let us now return to a strictly philosophical consideration: The name with which the beginning that has been found is named is not exclusively enclosed in the natural order of facts. All this is metaphysics, since the name infinitely overflows the fact. All this has its own logic for, if it is freedom, it is nevertheless not arbitrary; in fact, it finds correspondences and reasons. All this is, however, a metaphysics that, leaning out from the finite, finds the origin [of the finite] and prospects its end without being able to inhabit either [origin or end]. It instead nominates and, first of all, gives a name to the finite obstacle that it encounters by calling it "a being" [*ente*] and by ordering it within a horizon that is allegedly named "being" [*essere*]. It also encounters ancient names, which are decisive and fierily abstract. It teaches human beings how to handle anew the initial words; it learns the language again and perceives it again for the first time. It is not a spectator of the origin; rather, it stages it for the first time. The world of thinking is being born fresh and unexpected even in front of an old and sated world because it is not the consequential unfolding of something that was always forever known but rather is the risky coming across a first encounter. As Benjamin says with wisdom and depth in the fifth of the theses on the philosophy of history," "*das wahre Bild . . . huscht vorbei* [the true image . . . flits by]."[2] It does not simply pass nor does it simply fade away; rather, it slips away, goes away, in silence, tiptoeing, without being heard because that which is at stake, the essential, is also that which goes away, which fades away. The essential is threatened: it could disappear, it could even never happen. Philosophy encounters the beginning and the end without being able to occupy their positions. So as not to lose them or itself, it spells their name—in order to recall them. The name is invented by memory. [Memory] fixes a memory and orients the future according to it. The name recalls a meaning; yet, in uttering [the meaning], [the name] gives a sense orientation to such a meaning. Despite the force of such an orientation in the form of a project, which grants the possibility of an answer and the formulation of an appeal, the name does not ensure any domination whatsoever.

The name is the name of a difference—the encounter that is a beginning is also the beginning of a name. Jacob is no longer named Jacob but now Israel because of the hip that has been dislocated in the struggle and the encounter. The new name, the name of the beginning, carries the mark of a dislocation. If in a similar manner we considered also the name "being," the first philosophical name, not as the mark of a power or a transparency without residues but rather as the name through which we try to capture *das wahre Bild* of a reality that *huscht vorbei*, then would we not set out for a metaphysics that is not the tired and sterile repetition of an already made beginning, but rather a project that, albeit aware (to the point of the dislocation that runs through it) of the risks, risks the beginning and nourishes the hope of those who, although knowing they are entering an already initiated game, nevertheless do not want such a game to be already played out? Stated otherwise, [would we not set out for metaphysics as] a project in which the beginning is still always possible?

PRAISE OF PHILOSOPHY

Imagining that philosophy is the kingdom of concepts is to do it a wrong. Philosophy is rather the place of a tension: it is the urging of the specific absolute that is at one's heart more than anything else, and its unfolding in the form of universality and communication. Behind any great thought there is an absolute experience (or a fragmented series of absolute experiences) that comes to speech, gives itself to conceptual form, and subjects itself to the neutrality of the debate. It is easy to grasp how a tension is contained in all this: in the absolute experience, the experiencing subject and the thinking subject are one and the same; the one is steeped into the other in an immediacy preceding even speech articulation. In philosophical exhibition, the subject gives way to the objectivity of the matter that is being said, to the objectivity of the form of the argumentation, to the objectivity of the judgment and criticism that come from the outside. In this lies the great asceticism of the philosopher, in whom the urging speech subjects itself to the severe discipline of rules and structures of discourse that are not at the philosophers' disposal, but that ever since the beginning define the philosopher's possibility of reflecting.

Why such an ascetic discipline that trims the immediacy of speech? Why could we not rely on an immediate saying, in which the subjectivism of experience and speech are confessed once and for all with no masking and elusions? Why could we not rely on the simplicity of the one who, without reflection, nevertheless acts through gestures of life; or on the thickness of feelings of the one who fixes such gestures in an emotion?

There is no final answer to these questions except the millennium-long experience of a civilization that has preferred to entrust its most decisive expressions to the alterity of communicable language (to the point of risking the abstract neutrality of concepts) rather than to the pervasive immediacy of feelings. When it has not done so, such a civilization has experienced also the dangers of a speech that is transmitted by sympathy rather than through reflection, that arouses the connivance of shared sensations without the background of the severe and judging mediation of concepts.

Philosophy risks the loss of that which is most important for each—it sets out without having put anything in store—because it wishes to find it again as the most important for all. This is an immense claim, which nevertheless is the claim of any truth conquered in the arch that leads from "that which is important to me" to "that which is important to all."

To pursue this, philosophy invents worlds, constructs theories, and elaborates systems. But rather than encountering its object, all this, which wishes to protect life and stop it before it fades away, fails. Philosophy observes the birth of a world, [though only] when such a world is no longer being born and instead declines. Philosophy must recreate the possibility of such a birth, and nothing is more complex than constructing a simple object in a lab.

Philosophy saves the world because it makes possible for [the world] to be that which it is (sometimes and fragmentarily). When it makes life possible for others, philosophy itself dies, however, because one must indeed live and abandon philosophy again. Yet life that is simply restituted to itself withers; the simple becomes trivial. The sacrifice of thinking becomes necessary again.

Life and thought never proceed at the same pace; by failing each other, they come across each other. In this way, they redeem each other. The pain is indeed that this may not happen, for the same human being, at the same time.

Life has within itself the antidote for its own pain: it continues living indifferent to its own consumption. The pain of thought is greater because [thought] cannot discharge itself from that which it must yet cannot think, namely, life; because [thought] knows that only it itself could save life, and make life worth living. [Thought's] suffering is that of a doubly useless servant: because it does not fulfill its mission and because, if there is grace, the latter will restore with simplicity and without thought that which thought has sought for so long. Furthermore, [thought] is a rough and too sensitive servant; it perceives any dissent as if it were a newly received wound, yet it exposes itself to wounds and contests with fearless arguing. It is defenseless and armed, and in precarious balance between fulfillment and abandonment. While it pursues and even promotes differences, philosophy actually suffers from them as from a wound that is even more painful precisely when the agreement is proximate. The wound in fact discloses that my philosophy is not the truth (which, as long as one says it, is a sort of regulative supertruth; yet when one experiences it, it becomes a pain). Moreover, the proximity that suddenly diverges to the point of making one fear sidereal, infinite distances provokes a rift between me and the other who, by being proximate to me, essentially constitutes my identity.

In philosophizing, one could notice an entirely unjustified disproportion toward universality. Why would one risk the essentiality of that which matters in this dangerous everyone's land simply out of the desire of attaining it in the form of a concept? But is this disharmony not the properly human feature of doing philosophy? It is in fact akin to the disharmony that, within life, lets us grasp the absolute in the particular. In its own concreteness and identity, no love can sustain the absolute relevance that we assign to it; no ambition can live up to the universe of imagination that we connect to it. Nothing is ever adequate, yet we constantly live in the disproportion for the particular—without a certain woman, without the achievement of a certain goal, without a certain friendship we could not live; life would truly lose its meaning. And yet, for others, that very same friendship, that very same goal, that very same woman do not mean the same; they are valuable, they matter, but without being decisive.

It is not a surprise then that some thoughts, that is, the thoughts that think life, may become more important than life itself. It is another, yet analogous disproportion, which neither refutes nor closes the world of immediacy but rather opens a new world, which is neither less complex nor less intriguing.

In a sense, philosophy starts out on this path when, subtracting itself to the authority of the poet who fixes the particularity of a said into an eternal form, it agrees to become dialogue (words that are subjected to the wear and tear and measure of others). It immediately corrects itself, though, by formulating itself in a written dialogue (text, complex thread, where thin threads weave a strong fabric). Written dialogues give up the control and reassuring guarantee of self-correction; they formulate once and for all, they leave a written trace that cannot be deleted and is rather subjected to the disgrace of confutation, to the humiliation of not being understood and being rather rejected. This is a burning humiliation because behind well-constructed and solid argumentations full of references there is always the ajar shrine of an absolute word that had received welcome and is now threatened.

At its highest points, philosophy encounters madness again and again. It does so in order to exclude it. The ancient sages (partly prophets, partly poets) were in some way madmen, possessed by divine madness, as Plato claims in the *Ion*. Formalizing itself, philosophy has domesticated such latent madness and has turned it into an effort toward ascent. The imperative for thought has become to see, even to contemplate—in other words, to reduce the object to an immobility that lets itself be transfixed by the gaze of reason. Reason measures, that is, it excludes that which it does not succeed in reducing to its own size.

In the barely lit room in which modern philosophy modestly begins, Descartes grants doubt all possible variants. Yet from the start he excludes one such variant: that reason may be mad. Even when it posits the most critical beginning, the gesture of philosophy has always already excluded the total alterity of madness. Foucault is correct in emphasizing such a previous and irrevocable decision. But he completely misses his target because he does not realize that this is not a decision of philosophy; it is rather philosophy itself: reason constitutes itself by emerging from madness and casting it aside.[3]

For this reason the subject of modernity, which has lost certainty regarding its own reason, goes hunting for madness—in order to mirror itself in it and encounter reason anew. White, the color of immense and boundless madness, is the ghost of fear and hope; it is the feared and loved object that attracts and intercepts the life of any Captain Ahab. Like Melville's whale, madness has its smell, taste, and a language that fascinates. In order to be able to listen to it, one must fight it with the harpoon in one's hand or resort, like Ulysses, to the trick of being tied to the ship mast.

Philosophy is a solitary work; one risks no longer knowing whether one is sane or mad. Normal life is also not without its dangers; one might lose the sense of diversity, and no longer know whether one is mad or sane.

Reason, which has the power to exclude, nevertheless does not have the force to delete. In order to find itself, it must rather continually encounter the trajectories of madness—not in an adventure, while being intoxicated with what is timeless, but rather in the daily footsteps that trace and correct boundaries, that gain new land or abandon other land, that rejoin separate and noncommunicating islands and gain them to the present. The plough separates clods, cuts, divides. The project leans out over our madness and secures it to reason. In such a way, though, like the dark shadow of that which could have been (the halls that Cesare Pavese knows and adolescence must cross), madness also accompanies the days of our work; it is a defeated yet not extinguished, dangerous yet not useless power.

Philosophy prolongs the days of reason. Such days are numbered; they lack much of the divine madness, but they do preserve a spark of it, a spark that is already so blinding that often one has difficulties standing its sight. While losing madness, one has preserved a fragment of it.

CHAPTER EIGHT

Hermeneutics of the Positive

ALMOST A CONCLUSION

Were it not contradictory (philosophy is not only after the beginning, it is also always before the end), I would call this a conclusion. Certainly it is a conclusion in the sense of being a return home (yet, as always, the home has changed upon one's return). In this long journey we started out with time, with a praise of the present that wishes to consolidate philosophy as a meditation on time, and more specifically on the present time. We have then retraced the way of the tale as a form of making present within the conditions of finitude. Yet (and this is the place of the conclusion, the return home), philosophy is not a tale, a fable, or the invention of a subject that hands down its own legacy. Philosophy is the great tale of time. Not [simply] a tale but rather a great tale—that is, a tale that exhibits the very conditions of narrating. In this sense, it is a tale that founds the possibility of narrating. It founds it after having discovered (found [*trovata*]) it as the basis of all tales; and it founds it after having invented (found [*trovata*]) it as the condition of possibility of narrating. Philosophy is, therefore, not only a great tale, but rather the great tale of time; what it says is time, and narrating is a modality of the diction of time. Hence, the return: the present as departure and arrival point of our adventure.

POSITIVITY OF PHILOSOPHY

Before we close, a final relevant question remains for philosophy.[1] The fact that the conclusion of a philosophy is the opening of a new problem, and that the closing of discourse is also the new opening of thought, lets much of the ultimate, fascinatingly unconcluded nature of philosophy transpire. Let us return to the issue we wish to leave as a legacy for reflection.

What [reflection] says is (re)discovered [*(ri)trovato*]. Therefore, it precedes [reflection]. It is also realized [*trovato*] (posited). How is the relation constituted between what precedes philosophy (something that [philosophy] finds) and what philosophy introduces (that which it invents)?

This is a question that philosophy has explicitly asked itself in the last century in relation to the theme of its relationship with the positive that comes from religious revelation. As is well known, there have been two solutions: Hegel's and Schelling's. For Hegel, philosophy configures itself as the attempt at dialectically reabsorbing the revealed given within philosophical reason. In conformity with its general posture, thought seems to proceed according to a dual movement: On the one hand, it penetrates the essence of its own object; on the other hand, it is forced to represent such an object, and thereby to give it exteriority. This way, however, the object, which had been immediately penetrated, also appears as extraneous and therefore asks for a new understanding at a higher level. Accordingly, dialectics proceeds incessantly at least until the represented object contains all representable objects, and form and content reach full correspondence at last. For this reason, the ultimate object appears to be God, who is the supreme totality. For this, too, that is, in order that dialectics at last reaches its own rest, God must not appear as something that is found [*trovato*], as in religious revelation, but instead as something that is posited, as in the philosophical concept.

Both reason and religion are manifestations; both reason and religion are rationality. The revealed of religion that is incompatible with philosophy is the revealed as positivity that is found [*trovata*] because philosophy, which is divine service, posits and does not find its object. Revelation is overcome by philosophy not because of its contents (which in no way are the disclosure of something incomprehensible and obscure, but rather are the manifestation of some rationality) but because of its unresolved exteriority that dialectics brings to perfect mediation.

This is not the case with Schelling, who labels Hegelian philosophy as merely negative and abstract—an a priori empiricism or an apriorism [*apriorismo*] of the empirical. What ensues from this [namely, Hegel's philosophy] is the total misunderstanding of the value of revelation when taken in its autonomy and independence, and the absurd claim that philosophy produces its own objects.[2] Although not the starting point of philosophy, revelation (not unlike everything that is positive, such as existing nature, human being, and consciousness) confronts one with something that could not have been known except on the basis of the empirical fact of the revealed word (or existing nature, human being, or consciousness). Thus "revelation forces philosophy to go beyond merely logical systems."[3]

The confrontation on the theme of the positivity of philosophy has taken place in such authors with eminent reference to religious revelation, in the (already Hegelian) conviction that among all forms of positivity, [religious] positivity is the highest and the most solidly congenial to philosophy. All this, however, could not avoid raising the protest of theology, and more specifically that of the great revolutionary theologian Karl Barth. For him the issue is not that of understanding how thought posits itself in relation to revelation (whether it produces or interprets it), because in both cases a (necessary or free) solidarity and a possibility of understanding is presupposed. Revelation is rather such because it is unpredictable and paradoxical, ultimate and without appeal. For Barth, revelation is God's self-revelation; it precedes all dialectics.[4] The assertion that God reveals himself as the Lord is an analytical judgment, which shows that "God's revelation possesses its own truth within itself, and this is so both from an ontological and from a noetic standpoint." "According to the Scriptures, God's revelation is an absolutely fundamental fact that has its own reason only within itself and therefore represents, in relation to human beings, an ultimate and unappealable situation. Its reality and truth do not depend on any other reality or truth whatsoever. They cannot be judged, compared, or understood from out of any external criterion."[5]

Barth's position is completely overturned in comparison to the philosophical stance, which considers revelation as a given that is external to itself. It vindicates a rigid theological perspective according to which revelation is heard from out of a self-revealing subject and from out of the subject's action in the act of its own self-revealing.

It is both the same and not the same in subsequent theological developments. It is the same insofar as Bultmann, who remains loyal to the inspiration of *Zwischen den Zeiten*, also maintains that revelation is a kerygmatic event, a divine initiative, an act that I do not encounter but by which I am encountered and involved. It is not the same, because understanding such a paradoxical revelation becomes an imperative one cannot avoid. Revelation is the unveiling not of something mysterious but rather of life—of the life that is Christ. Such a revelation does not communicate a form of knowledge in an Enlightenment style; rather, it makes one be part of an event. Nevertheless, we cannot avoid asking whether and how such an annunciation affects and involves us.

> The revelation is something that addresses us (*sie redet uns an*). That human beings learn to understand themselves in it means that *they learn to understand each of their Now, every instant as qualified through revelation* . . . What has been revealed, then? Absolutely nothing insofar as, in revelation, one looks for teachings to which no human being could ever arrive, for secrets that, once they have been communicated, are learnt once and forever. Yet everything has been

revealed insofar as revelation enlightens human beings on themselves in such a way that they can newly understand themselves.[6]

The issue of understanding is solved through a circle, which is the circle of an existential hermeneutics. Barth's vindication, not without the Basel theologian's surprise and misunderstanding, bends again toward the human being and with Bultmann attests the inextricability of divine and human entanglement.

The theological protest refutes neither Hegel nor Schelling. Simply, as almost always happens in the vicissitudes of thought, it displaces them differently and perhaps contributes to put in brackets the too direct confrontation between philosophy and revealed given. What is certain is that twentieth-century philosophy moves along other paths, and has left aside the confrontation between reason and revelation.

Heidegger shifts the dimension of revelation back to truth (and no longer to the Scriptures or to Reason). "What does it mean 'there is truth'? 'We' presupposes truth because, 'we,' existing in the kind of being of Da-sein, *are* 'in the truth' . . . We do not presuppose truth, but *truth* makes it ontologically possible that we can *be* in such a way that we 'presuppose' something. Truth first *makes possible* something like presupposition."[7] Heidegger's itinerary grafts disclosedness onto truth and the heart of being in a radical way that grounds revelation in elementary existential words. The salient element of his path lies in the move that turns truth into an act of exiting concealment or disclosedness: revelation is once again surprised in the act of its self-disclosing. Truth is unveiling, the exit from the concealment that equally belongs to it because of its being-in-the-world. Revelation is not manifestation but rather an act of self-manifesting, existence. Here Heidegger's philosophy shows its deeply phenomenological feature, for which the description of the structures of existence is the ability to surprise the thing in its own self-constitution. We have a confirmation of this in the letter to Jean Beaufret accompanying the "Letter on Humanism," in which Heidegger writes: "Dasein is a fundamental word in my thinking, and therefore it is also the occasion for great misunderstandings. For me, Dasein does not mean *me voilà* as much as, if I can say it in a perhaps impossible French, *être le-là*. And *le-là* is the same as truth—un-hiddenness-opening."[8]

Heidegger's move goes then even further (or perhaps, on the contrary, it goes back toward a more remote and forgotten root): the revelatory dimension belongs to truth as the most originary moment of them all. Heidegger's (and in general, the hermeneutic) attaining of truth as revelation that exits concealment must compensate for the difficulty of reaching judgment and discerning false from true. The truth that is in the world, and that is truth intersecting and nourishing existence, finds itself as ambiguously situated in truth and untruth, in unconcealment and concealment. As a consequence, revelation also ultimately concerns in the same way both the truth and the untruth of the truth. Heidegger's radical move, which turns revelation into the way in which being gives itself, constitutes a sort of secularization of theology, whose contents are transcribed/assumed according to a format that in many aspects retrieves, without dialectics, Hegel's mode of philosophy.

At the end of the journey we find ourselves, paradoxically, once again very close to Hegel. Revelation, that is, the given, turns out being the posited or, after the theological protest, the self-positing, the activity of positing—that which constitutes being in the most intimate and secluded manner. Without resorting any longer to the necessity of the movement of thought, we are nevertheless in front of a descriptive unfolding that highlights an essential movement: both unveiling and being veiled belong to the truth, and such an ambiguous co-belonging is disclosed in the act of unveiling. Thought is revelation that unveils itself without severing its own ties from belonging to the dimension of forgetfulness of this world.

PRAISE OF DANGEROUS MEDIATION

What is required of contemporary philosophy is precisely the (maieutic and unhappy) location that in Chaim Potok's novels is entrusted to Reuven and his father. I am thinking especially of *The Chosen* and *The Promise*.[9] Reuven and his father appear as isolated characters, somewhat inscribed in a circle of misunderstanding. They nevertheless constitute the bridge between Orthodox Jews, afraid of losing even just one fragment of their patrimony, and those who are fond of the tradition yet no longer believe in it (as happens for the character of Gordon). Either of these [groups] is too weak to withhold a good that is threatened. In order to preserve the tradition, one must work critically against it (and this is dangerous, in the sense that this offers weapons such that the weapons themselves may threaten and endanger the tradition). The problem of continuity arises out of a discontinuity that is no longer disguised. Such a work is by definition always exposed to risk, and can only rely on seriousness of intentions and blamelessness of behavior. Yet anyone, on either side, can blame [such work] for its being too open to compromises.

A philosophy that is willing to join contrasting reasons must defend itself from the very same accusation. Does one not need courage, firmness, labor, and suffering so as to hold on to a form of wisdom that is not simply the shortcut of a brilliant and successful radicalism? Does one not also need a discreet humility, exactly as it happens to Potok's characters, who are never at the center of action and attention although they are the engine and secret power of attraction?

The author who provides us with the major elements for such a dangerous mediation is Luigi Pareyson, whose [notion of] interpretation appears as the proper mode that fallen existence has so as to relate itself to truth as revelation. Truth is indeed unconcealment, and hermeneutics is indeed maieutics of unconcealment, yet philosophical reason is not immediately the place of truth (and neither is existence). We are here confronted with a perspective that operates not through the confutation of previous claims, but rather through a sort of side step, a change in angle, a gesture of freedom. The concretion of truth is already there. Yet such an already-there is not the human being but rather something (revelation? Or the historical forms of revelation? Or myths? Or art? Or common sense?) to which to turn/return [*ri-volgersi*].

Between truth and interpretation there is reciprocal correspondence and even originary solidarity, but also radical dissymmetry. Although it gives itself in interpretation, truth is in fact not exhausted within it because precisely the inexhaustibility of truth, which maintains its own irrepressible transcendence, is at the ground of the multiplicity of interpretations. Interpretation is the self-giving of truth within finite (one could say fallen) existence. "Interpretation is a type of knowledge intimately constituted by the constant risk of failure, in which revelation is obtained only as victory over the constantly current threat of concealment. Such precariousness in interpretation is due to the alternative freedom itself poses to the person."[10]

A theme thus appears that, although undeveloped in [Pareyson's magnus opus] *Verità e interpretazione*, nevertheless will become central in subsequent works and will powerfully condition Pareyson's tough coming to terms with Heidegger. It is the theme of evil, or fallenness. Against Heidegger's approach, which does not choose, which perceives the ambiguity of truth and untruth yet does not give untruth a name, Pareyson begins to build elements of his own self-distancing [from Heidegger] by resorting to the centrality of the theme of interpretation, which progressively takes up non-Heideggerian cadences. It is in the already mentioned inaugural lecture at the international seminar devoted to Heidegger at the Istituto Universitario Suor Orsola Benincasa in Naples, March 13–17, 1989,[11] that Pareyson simultaneously comes to terms with Heidegger and with himself—if it is true, as he remarks, that up to that point he had been unable to take a critical position toward Heidegger because of his own excessive involvement with such an author's thought. The distancing from Heidegger must, therefore, also be understood as self-critical distancing.

This occurs first of all through a harsh yet historiographically correct judgment regarding Heidegger's anti-Christianity. For a thinker like Pareyson, who in those years was engaged in a hermeneutics of Christianity, such a judgment does not have an exclusively historiographic value.

Rather, it implies the refusal of a philosophical approach that in truth had belonged to himself too. Secondly, [the distancing] occurs through the accusation moved to Heidegger's philosophy of being a "philosophy of ambiguity" and "a blocked radicalism" because it is unable "to grasp the principle of ambiguity that can only be freedom, on which the whole of reality hangs."[12] For this reason, anticipating subsequent developments and making explicit the intuitions on which he had meditated for long, Pareyson concludes by saying that it is necessary "to abandon the centrality of being and replace being with freedom."[13]

Christianity alone could have suggested to Heidegger the centrality of freedom as faculty of good and evil. Like the entire modern philosophy, Heidegger does not, however, ask the question of evil in the proper manner. Indeed, Heidegger speaks of the negative, negation, and negativity to the point of characterizing being itself as nothingness. Yet, such [a conception of] nothingness essentially means "the reserve of being so as not to be confused with beings, and not a negativity such as that of evil, which is negation, true and real destructiveness."[14]

Revelation and truth revert to a dissymmetry between the act of revealing, which belongs to thought, and that which is revealed, which belongs to the order of truth. By intertwining the theme of interpretation, Pareyson's themes of evil, sin, and fall produce unexpected effects, and dramatize the crystalline (and ultimately neutral) transparency of Heidegger's revelation, which, having reabsorbed the originary theological value within itself, turns into simple disclosedness. They reopen the play of thought through their remarking an unexpected dissymmetry. They thus invite us to retrieve a question that might have appeared exhausted.

From this perspective, the dialectics between found [*trovato*] and posited becomes less extrinsic, and the opposition between revelation and reason appears less rigid. In the end, the whole vicissitude we have been following is a confirmation of this. Leaving what specifically belongs to the devised solutions aside, is what is most true in what we have observed not the co-belonging of the two moments (Hegel and Schelling) and, moreover, the reassessment within revelation of reason (which in Barth takes up the form of self-understanding and in Bultmann the form of the nexus between word and existence), and within philosophical truth of an intrinsically revelatory dimension (Heidegger)? In this way, is not a nexus indicated that connects posited and found, rational and positive, revelation and understanding? Is the co-belonging of the terms not suggested? Should one then perhaps make a last move, which Pareyson indicates to us—namely, that of exploring the nexus not in the form of identity and symmetry, but rather in the form of difference and dissymmetry? What is revealed is not what is thought, yet what is thought is not without what is revealed. The positive is not the posited, yet the posited is not without the positive. Religion ([which is an] eminent symbol of a positive that exists without having thought as its own origin; and one could say the same for art, history, experience, or life) is not philosophy; yet, philosophy is not without the relation with what is other than itself, which it must think.

One thus returns to religion as posited, to a reserve that is not to one's avail. Moreover, and even more decisively, a dissymmetry is reinstituted between the act of revealing that belongs to thought and that which reveals itself, which belongs to the order of truth. The disjointure that Pareyson introduces between truth and interpretation lets one glimpse a stigma that challenges anew (but also justifies) the whole itinerary we have followed.

On the one hand, philosophy cannot in fact accept anything sacred, cannot tolerate any reserved domain, and, in a Hegelian mood, must translate that which has been found into that which has been posited; it must deny positivity in order to think it. This is the inevitable movement of thought. Thought making such movement, on the other hand, cannot not realize that it does not find its own content as an immediate self-production. In the labor of the concept, it generates it as a mediated work of production, which brings to thought some existent that is antecedent to thought itself. Philosophy's task is to understand that beyond itself is a found [*trovato*] that it itself has not posited, although it can only think of [such found] by positing it as an object of thought. That which is posited as an object of thought, however, is not the found that antecedes thought; it is a specific found, which has been found again, re-found through a double movement, that is, as encountered and as imagined—imagined because encountered (at least in the obtuseness of object resistance) and encountered because imagined (at least by thought's encompassing will).

In this sense, philosophy does not finally come to understand that which had already been there; it is not the rational understanding of the existent, which is generated as object for thought. Rather differently, philosophy is the human work of finitude that, within the present, discriminates between an inert found, which is encountered, and an abstract posited, which is projected. [Philosophy] joins the found as that which is projected and the encountered as that which is real; it unfolds them in the form of a tale; it gives them the consistency of a temporal order between the times of the past and of the future. Between the times of the eternal and of the existent, between the pillars of an ontological proof and of the *cogito*, it generates its own temporality, the divide of the present that exists only as opening of a world and sense, as argued and narrated distension of a threshold.

The present is the time of philosophy because philosophy interrupts mere flowing in order to speak instead of a possible consistency, which is naturally simply posited—a choice, a proposal. It is nevertheless a meaningful proposal, which enables one to find that which would otherwise go lost or would not come to existence.

At the end of our tale, philosophy configures itself as hermeneutics of the given, of the positive as truth preceding but also following philosophy; as hermeneutics within the in-between, which clings to and unfolds the positive. Consequently, [philosophy configures itself] as hermeneutics of the present time, as finite unfolding of the infinite, as prolonging of the instant in the form of a consistent duration. This can be done only by a philosophy that is not knowledge but rather love of knowledge—that is, a finite knowledge, a knowledge of the finite, a fissure that interrupts meaninglessness and inaugurates a possible present.

Notes

PREFACE

1. Paul Celan, "Cello Entry," from "Atemwende," in *Poems of Paul Celan*, trans. Michael Hamburger (New York: Persea Books, 1988).

2. Two essays especially functioned as preparation: "Il racconto della filosofia," in *Annuario Filosofico* 14 (Milan: Mursia, 1998), 61–98; and "L'intersoggetività del raccontare," in *Intersubjectivité et théologie philosophique*, ed. Marco Maria Olivetti (Padua: Cedam, 2001), 411–29. These are texts that, albeit retrieved, have been inserted in the context of a further reflection, and thus they have been even more deeply modified. The new context has allowed also for the retrieval of some passages from "Al limite del finito," in *Ermeneutiche della finitezza*, ed. Giovanni Ferretti (Pisa-Rome: Istituti editoriali e poligrafici internazionali, 1998), 207–16, and from *Le passioni del finito* (Bologna: Dehoniane, 1994).

CHAPTER ONE — DRAMATURGY OF THOUGHT

1. My attention has been drawn to this point by Sergio Rostagno, *Karl Barth* (Brescia: Morcelliana, 2003), 18. In particular, the second section devoted to the conception of time (ibid., 25–33) shows very convincingly how Karl Barth, using the reflections by his philosopher brother Heinrich, arrives at conceiving of "time as of the human participation to the origin, a participation that is made possible, recommended and guided as recognizing/grateful act" (28). When translating into the language that I am here adopting, the temporality of the present is not thought of from out of itself; nevertheless, when considered from the perspective of the origin, it is secured precisely on behalf of theology.

2. [Trans. note: This is the title of the journal started in 1922 by Barth, Thurneysen, Gogarten, and Metz.]

3. Karl Barth, *The Epistle to the Romans*, trans. Edwyn C. Hoskyns (New York: Oxford University Press, 1965), 149.

4. Ibid., 304.

5. Ibid., 305.

6. Ibid., 49.

7. Ibid., 47.

8. Ibid., 116. [Trans. note: In Hoskyns's English translation, what has been rendered here as "contemporaneousness" appears as "likeness."]

9. Ibid., 141.

10. In this section I make use of suggestions taken from a seminar I held with the students with whom for many years I have had the luck to share various research projects. More specifically, in more than one place I have availed myself of what was argued by Iolanda Poma. I would like to refer also to Iolanda Poma, *Eresie della fenomenologia* (Naples: ESI, 1996).

11. See Edmund Husserl, *The Phenomenology of Internal Time-Consciousness*, trans. James Churchill (Bloomington: Indiana University Press, 1966), 52.

12. Ibid., 43–44.

13. So much so that, when speaking of recollection (the argumentative mechanism can however be applied more extensively), Husserl can write that "the temporal present in recollection is remembered, presentified. And the past is remembered in the same way, presentified but not perceived. It is not the primarily given and intuited past. On the other hand, the recollection itself is present" (ibid., 58).

14. For the first aspect, see ibid., chapter 1; for the second, see ibid., 104. The fact that, as well explained in an essay by Bianchin, Husserl's reflection on time could not avoid having consequences for subjectivity to the point of posing the question of its temporalization is a matter that, albeit important, is of no concern here. See Matteo Bianchin, "Husserl, dalla percezione del tempo alla temporalizzazione della soggettività," in *Tempo e interpretazione*, ed. Luigi Perissinotto and Mario Ruggenini (Milan: Guerini, 2002), 215–29.

15. Martin Heidegger, *Being and Time*, trans. Joan Stambaugh (Albany: SUNY Press, 1996), 304.

16. Ibid., 22.

17. Ibid., 405.

18. Ibid., 304.

19. Ibid., 389.

20. See Ugo Perone, *Modernità e memoria* (Turin: SEI, 1987), 66–69.

21. See the chapter devoted to Derrida in Emmanuel Levinas, *Proper Names*, trans. Michael B. Smith (Stanford: Stanford University Press, 1996), 55–62.

22. See Jean Greisch, *Ontologie et temporalité. Esquisse d'une interprétation intégrale de* Sein und Zeit (Paris: PUF, 1994).

23. Luigi Pareyson, "Heidegger: la libertà e il nulla," in *Annuario filosofico* 5 (Milan: Mursia: 1989), 9–29.

24. Walter Benjamin, *The Arcades Project*, trans. Howard Eiland and Kevin McLaughlin (Cambridge, MA: Harvard University Press, 1999), 471.

25. I would like to pass over further considerations that would complete the picture and delineate the alternative between an ecclesiology of the institutions and the mere radicalism of transcendence. It will suffice here to highlight the evident parallelism in the situation, which makes one anticipate also a parallelism as for the outcomes.

CHAPTER TWO — THE PRESENT AS THRESHOLD

1. Siegfried Lenz, *The German Lesson*, trans. Ernst Kaiser and Eithne Wilkins (New York: Hill and Wang, 1971), 344.

2. Ibid.

3. This theme is repeatedly present in my reflection and will reappear in many places in the present work too. For a more systematic albeit brief treatment of the historical alternatives to which I am referring, see Ugo Perone, "Limite e soglia," *Filosofia e teologia* 1 (2004): 7–15.

4. The concept of "threshold" (*Schwelle*) is unfortunately not (and I think this is a mistake) among those encompassed in the two however precious volumes devoted to *Benjamins Begriffe*, ed. M. Opitz and E. Wizisla (Frankfurt am Main: Suhrkamp, 2000). One can only refer then to W. Menninghaus, *Schwellekund. Walter Benjamins Passage des Mythos* (Frankfurt a. M.: Suhrkamp, 1986).

5. Benjamin, *Arcades Project*, 87.

6. Ibid., 88.

7. Walter Benjamin, "Berlin Childhood around 1900," in *Selected Writings*, ed. Howard Eiland and Michael Jennings (Cambridge, MA: Harvard University Press, 2006), vol. 3, 396.

8. It would be interesting to follow the role of the corporeal in modern philosophy, but this would lead us too far from what we wish to establish here. I only note that there are specific sites where the slidings of which we are talking here get deposited, almost unaware. To state briefly: Merleau-Ponty's turn toward the flesh as well as his denunciation of the impossibility of the Husserlian reduction inherit and continue the intention of phenomenology of producing a rigorous science that, according to Husserl's explicit intention, brings to fulfillment the ideal proper to classical as well as Christian philosophy. Merleau-Ponty replaces the terms that, in the spirit of such a tradition, are still at work in Husserl—the subject and its interiority on the one side and the truth on the other—with a "being in the world" where the subject is our body and the truth is its belonging to space and time. In sum, the body in the world inherits the metaphysical function of the traditional subject and interiority, and with them it also inherits their aporias. See Maurice Merleau-Ponty, *Phenomenology of Perception*, trans. Colin Smith (New York: Routledge, 1996), XI, which is an explicit translation of Augustine that is however implicit in Husserl, and more precisely in the conclusive sentence of his *Cartesian Meditations.*

9. See, in the present chapter, the section titled "The Present Is Not Presence but Divide, Condition of Temporality."

CHAPTER THREE — ETHICS OF THE PRESENT

1. See Alessandro Manzoni, *The Betrothed*, trans. Bruce Penman (New York: Penguin, 1984).

2. See Remo Bodei, *Geometria delle passioni* (Milan: Feltrinelli, 1991).

3. See Paul Hazard, *The European Mind, 1680–1715: The Critical Years*, trans. J. Lewis May (New York: Fordham University Press, 1990), 299.

4. 1 Thess. 4:10–12.

5. Jean-Luc Nancy, *The Experience of Freedom*, trans. Peter Fenves (Stanford: Stanford University Press, 1994), 146–47.

6. Benjamin, "Berlin Childhood," 357.

7. Merleau-Ponty, *Phenomenology of Perception*, 197.

CHAPTER FOUR — TALE WITHOUT AUTHOR

1. Pierre Janet, *L'évolution de la mémoire et de la notion du temp* (Paris: Chahine, 1928), 308–9. I owe the quotation to the interesting book by L. Sozzi, *Vivere nel presente. Un aspetto della visione del tempo nella cultura occidentale* (Bologna: Il Mulino, 2004), which unfortunately I happened to read too late, but which includes an abundant wealth of considerations about the present. With a magisterial virtuosity Sozzi spaces among various national literatures and philosophy and, without taking a final stand, nevertheless shows the richness, variants, and tensions of the theme of the present. In this regard it certainly converges with my attempt, since it illustrates with plenty of references the centrality of the present in any vision of temporality. The quotations from Janet are found in Sozzi, pp. 272–73, and I quote them in Sozzi's rendition.

2. On all this, see what is said by Walter Benjamin, "On Language as Such and on the Language of Man," in *Selected Writings*, vol. 1, 62–74, and "The Task of the Translator," ibid., 253–63.

3. On this, see the section titled "Finite Existence" in Chapter Six of the present book.

4. Anselm, "*Proslogion*, Preface," in *The Major Works*, ed. Brian Davies and G. R. Evans (New York: Oxford University Press, 1998).

5. Ibid., chap. 5.

6. Ibid., chap. 15.

7. Francesco Tomatis, *L'argomento ontologico. L'esistenza di Dio da Anselmo a Schelling* (Rome: Città Nuova, 1997) offers all the elements for a reconstruction of the history of the ontological argument. I would also like to refer to some classical texts such as Karl Barth, *Anselm: Fides Quaerens Intellectum* (New York: Meridian Books, 1962); Dieter Henrich, *Der ontologische Gottesbeweis* (Tübingen: Mohr, 1960); W. Röd, *Der Gott der reinen Vernunft. Die Auseinandersetzung um den ontologischen Gottesbeweis von Anselm bis Hegel* (Munich: Beck, 1992).

8. The interpretation given by J. Ringleben, *Erfahrung Gottes im Denken. Zu einer neuen Lesart des Anselmschen Argumentes (Proslogion 2–4)* (Göttingen: Vandenhoeck und Ruprecht, 2000) rotates around the thesis that for Anselm "thought is truly thought only as thought concerning God" (ibid., 33), and that giving up thinking about God implies giving up thought altogether. Although I move from a different perspective, there are interesting correspondences and confirmations with respect to what I have said that transforms the ontological proof into, as it were, the superior margin of an ontological need for the coincidence of existence and essence. This is also confirmed by the relevance that the theme of the "*certe* [truly]," which I have discussed, has in Ringleben too.

CHAPTER FIVE — THE TALE OF THE I

1. See Martial Gueroult, *Descartes' Philosophy Interpreted According to the Order of Reasons*, trans. Roger Ariew, 2 vols. (Minneapolis: University of Minnesota Press, 1984).

2. See Michel Foucault, *The Order of Things: An Archaeology of the Human Sciences* (New York: Random House, 1970).

3. Doubt leads to despair, as is so familiar to nineteenth-century philosophy, for which *zweifeln* [doubting] is immediately *verzweifeln* [despairing]. See Hegel's exemplary remark, which does not go in the direction Kierkegaard will then take in "*De omnibus dubitandum est*," but which already shows how Descartes' radicalness is not the ultimate radicalness but rather a temporary situation. See Georg Wilhelm Friedrich Hegel, *Phenomenology of Spirit*, trans. A. V. Miller (New York: Oxford University Press, 1977), 49.

4. See Gershom Scholem, *Major Trends in Jewish Mysticism* (New York: Schocken Books, 1941), 349–50.

5. The figure of Celan is to be inscribed with full rights within this philosophical path, whose moves come from far away, from Anselm's transparencies to arrive at modernity. Not only has Celan frequently cited philosophers, as is shown by his readings of Buber, Adorno, Benjamin, and especially Heidegger, but also the philosophers have frequently cited Celan, as is shown by the writings of Levinas, Gadamer, Derrida, Pöggeler, Nancy, and Bollack, among others. The list could go on so far that a study of Celan and the philosophers seems to me to be necessary, because it could perhaps help in the understanding of both Celan's arduous saying and of the *proprium* belonging to philosophy.

In the following paragraphs my writing, nourished by Celan's images, will necessarily become more allusive and, in the spirit of Celan, can be read according to more than one register. This is according to Celan's own mode; speaking of Büchner who spoke of Lenz, Celan spoke of himself. And Lenz's January 20 is the January 20 of the Wannsee decision of the final liquidation of the Jews. To speak of oneself, one needs to speak of others. Even, the reversal that allows us moderns not to entirely lose our connection with the ancients is the detection, in such a tough law, of a possible salvation and the acquisition of a never-heard-of conquest: my own self-estrangement—the upside-down I who walks on its head. The certainty of the Cartesian I, proud of itself, appears to be regained in other ways, after one had simply set it on the side, without even confuting it, as occurs in the most painful departures. The upside-down I is grasped through oblique investigations; they are oblique like the speech that must open its way while proceeding masked (yet Descartes, like all great thinkers, had already foreseen even this, even against himself: *larvatus prodeo* [I proceed in disguise]).

6. Among Celan's very few and precious writings in prose, the most important as document of this author's poetics is "The Meridian," a speech he delivered on October 22, 1960, on the occasion of the awarding of the Büchner Prize in Darmstadt. Moving from Büchner's Lenz (which is also, as is customary, a way of paying homage to the one from whom the prize takes its name), who had told the tale of the tragic irruption of madness into Lenz's life (as a person, not as an artist) on some January 20, Celan unfolds his own meditation—in an air that has

become unbreathable—on the meaning of dates in poetry, on poetry's inclination to become silent, on the impossibility of finding the places of one's childhood on the map, and on a paradoxical utopia. "The Meridian," translated by Jerry Glenn, is reprinted as an Appendix in Jacques Derrida, *Sovereignities in Question: The Poetics of Paul Celan*, ed. Thomas Dutoit and Outi Pasanen (New York: Fordham University Press, 2005), 173–85. All quotations are from this text.

7. Celan, "The Meridian," 175.

8. Celan, "Conversation in the Mountains," in *Collected Prose*, trans. Rosemarie Waldrop (New York: Routledge, 2003), 22. [Trans. note: The remarks in brackets are Perone's.]

9. *Die Niemandsrose* is the title of a 1963 collection of poems. The expression recurs in "Psalm," where Celan evokes the paradoxical blooming of a creatural Nothingness out of love for a metaphysical No one towards and against (*Dir/entgegen*) whom the rose blooms: "No one moulds us again out of earth and clay, / no one conjures our dust. / No one. // Praised be your name, no one. / For your sake / we shall flower. / Towards you. // A nothing / we were, are, shall / remain, flowering: / the nothing-, the / no one's rose"; in *Poems of Paul Celan*, 175.

10. Celan, "Mandorla," from *Niemandsrose*, in ibid., 189. The locution *steht und steht* [dwells and dwells] is repeated three times in fifteen verses. The verb *steht* [dwells] appears altogether fourteen times.

11. Karl Löwith, *Gott, Mensch und Welt in der Metaphysik von Descartes bis zu Nietzsche* (Göttingen: Vandenhoeck und Ruprecht, 1967).

12. Celan, "Zürich, the Stork Inn," from *Niemandsrose*, 157: "Of your God was our talk, I spoke / against him, I / let the heart that I had / hope: / for / his heighest, death-rattled, his / quarreling word - / Your eye looked on, looked away, / your mouth / spoke its way to the eye, and I heard: / We / don't know, you know, / we / don't know, do we?, / what / counts."

13. Celan, "Corona," from "*Mohn und Gedächtnis*," in *Poems of Paul Celan*, 59: "My eye moves down the sex of my loved one: / we look at each other, / we exchange dark words, / we love each other like poppy and recollection, / we sleep like wine in the conches, / like the sea in the moon's blood ray. // We stand by the window embracing, and people look up from the street: / it is time they knew! It is time the stone made an effort to flower, / time unrest had a beating heart. / It is time it were time. // It is time."

14. Celan, "The Meridian," 175.

15. The now [*l'oggi*] finds itself precisely in the middle, and works as a discrimination in two important poems, devoted respectively to Heidegger and Hölderlin, titled "Todtnauberg" from "*Lichtzwang*," in *Poems of Paul Celan*, 293–94 (this was written after the visit to the famous *Hütte* in the Black Forest) and "Tübingen, January," from *Niemandsrose*, ibid., 177.

16. See Celan, "The Meridian," 175–76.

17. Ibid., 178.

18. Ibid.

19. Ibid., 180. [Trans. note: The translation has been modified to reproduce Perone's text, which has "*poesia* (poetry)" where the English translation reads "literature."]

20. Ibid.

21. Ibid., 180–81. [Trans. note: The translation has been modified to reproduce Perone's text, which has "*si affida* (entrusts itself)" where the English translation reads "addresses."]

22. Ibid., 181.

23. Ibid., 182.

24. Paul Celan, *Gesammelte Werke*, ed. Beda Allemann and Stefan Reichert (Frankfurt am Main: Suhrkamp, 1983), vol. II, 312.

25. Celan, "The Meridian," 179.

26. Ibid., 185.

27. Ibid., 183.

28. Ibid., 185.

29. Celan, "Psalm," in *Niemandsrose*, 175.

CHAPTER SIX — THE TALE OF FINITUDE

1. [Trans. note: *Holiday* in the sense of what is not a weekday or workday, and not in the sense of days of vacation.]

2. See Luigi Pareyson, "Frammenti dell'ontologia," in *Ontologia della libertà* (Turin: Einaudi, 1995), 293–345.

3. Perone, *Nonostante il soggetto.*

4. *The Arabian Nights: Tales of 1001 Nights*, vol. 3, trans. Malcolm C. Lyons (New York: Penguin, 2010), 733.

5. Ibid.

6. Walter Benjamin, "The Storyteller," in *Selected Writings*, vol. 3, 151.

7. *The Arabian Nights*, 734.

8. Ibid., 733.

9. Ovid, *Metamorphoses*, trans. Frank J. Miller (Cambridge, MA: Harvard University Press, 1966), bk. III, v. 361.

10. Ibid., bk. III, vv. 380–82: "*Dixerat: 'Ecquis adest?' et 'adest' responderat Echo: / Hic stupet, utque aciem partes dimittit in omnis, / Voce 'veni' magna clamat; vocat illa vocantem*"; "[he] had cried: 'Is anyone here?' and 'Here!' cried Echo back. Amazed he looks around in all directions and with loud voice cries: 'Come!'; and 'Come!' she calls him calling."

11. Aristotle, *Metaphysics*, trans. Hugh Tredennick (Cambridge, MA: Harvard University Press, 1933), 1006a.

CHAPTER SEVEN — THE GREAT TALE OF TIME

1. Alessando Manzoni, "Marzo 1821." [Trans. note: This is the title of one of Manzoni's most inflammatory patriotic poems, whose conclusive lines are here cited by Perone.]

2. Walter Benjamin, "On the Concept of History," in *Selected Writings*, vol. 4, 390.

3. On this topic see Pier Aldo Rovatti, *La posta in gioco. Heidegger, Husserl, il soggetto* (Milan: Bompiani, 1987), especially Chapter 4 in the first part, which

sums up the salient moments of the debate between Foucault and Derrida, a debate that, starting with Descartes, takes up a much wider relevance and implies the very issues I have mentioned here.

CHAPTER EIGHT — HERMENEUTICS OF THE POSITIVE

1. I have dwelled on these issues more analytically in Ugo Perone, "La rivelazione e il pensiero filosofico del Novecento," in *Il prisma della rivelazione. Una nozione alla prova di religioni e saperi*, ed. S. Sorrentino (Assisi: Cittadella, 2003), 85–100. More precisely than what was already indicated in that essay, today I would consider it necessary to proceed beyond Heidegger with a hermeneutics that, by understanding the reasons of the philosophical resorting to the concept of revelation/revealedness, is capable of foreclosing all forms of encompassing and incorporation and thinks more radically from out of a nonbridgeable difference.

2. See Friedrich Wilhelm Joseph Schelling, *Sämtliche Werke*, ed. M. Schröter (Munich: Beck and Oldenbourg, 1927–1943), VI E, 396ff.

3. Ibid., 138. One should remark, however, that for Schelling, beside revelation there is "an internal necessity within philosophy itself" that pushes in such direction (ibid.).

4. See Rostagno, *Karl Barth*, 41, 46. Rostagno correctly emphasizes, as a confirmation of the theological radicalness of the concept of revelation, how the theme of the Trinity is not the fulfillment but rather the foundation in Barth's dogmatics.

5. Karl Barth, *Dogmatique* (Geneva: Labor et Fides, 1953), vol. II, 11, 24.

6. Rudolf Bultmann, *Glauben und Verstehen* (Tübingen: Mohr, 1933–1965), vol. III, 21, 29–31.

7. Heidegger, *Being and Time*, 209.

8. See Martin Heidegger, "Lettre à Monsieur Beaufret," in *Lettre sur l'humanisme*, trans. R. Munier (Paris: Aubier, 1964), 182–84.

9. Chaim Potok, *The Chosen* (New York: Simon and Schuster, 1967) and *The Promise* (New York: Alfred Knopf, 1969).

10. Pareyson, *Verità e interpretazione*, 84. [Trans. note: A translation of this work by Pareyson is currently being carried out by Robert Valgenti for SUNY Press.]

11. The title of the lecture was "Nothingness and Freedom as Beginning"; it was published with the title *Heidegger: la libertà e il nulla*.

12. Ibid., 22.

13. Ibid., 25. I can attest how, ever since my first thesis work, *La filosofia della libertà in Charles Secrétan*, later published as memoir of the Academy of Sciences in Turin in 1968, Pareyson displayed not only a very sharp interest for the primacy of freedom, but also for the ontological meaning of such a primacy.

14. See Pareyson, *Heidegger: la libertà e il nulla*, 27.

Bibliography

Anselm of Canterbury. *The Major Works.* Edited by Brian Davies and G. R. Evans. New York: Oxford University Press, 1998.

Arabian Nights: Tales of 1001 Nights. Translated by Malcolm C. Lyons. New York: Penguin, 2010.

Aristotle. *Metaphysics.* Translated by Hugh Tredennick. Cambridge, MA: Harvard University Press, 1933.

Barth, Karl. *Anselm: Fides Quaerens Intellectum.* New York: Meridian Books, 1962.

———. *Dogmatique.* Geneva: Labor et Fides, 1953.

———. *The Epistle to the Romans.* Translated by Edwyn C. Hoskyns. New York: Oxford University Press, 1965.

Benjamin, Walter. *The Arcades Project.* Translated by Howard Eiland and Kevin McLaughlin. Cambridge, MA: Harvard University Press, 1999.

———. *Selected Writings.* Edited by Howard Eiland and Michael Jennings. Cambridge, MA: Harvard University Press, 2006.

Bianchin, Matteo. "Husserl, dalla percezione del tempo alla temporalizzazione della soggettività." In *Tempo e interpretazione*, edited by Luigi Perissinotto and Mario Ruggenini, 215–29. Milan: Guerini, 2002.

Boccaccio, Giovanni. *The Decameron.* Translated by Mark Musa and Thomas G. Bergin. New York: Signet Classic, 2022.

Bodei, Remo. *Geometria delle passion.* Milan: Feltrinelli, 1991.

Bultmann, Rudolf. *Glauben und Verstehen.* Tübingen: Mohr, 1933–1965.

Celan, Paul. *Collected Prose.* Translated by Rosemarie Waldrop. New York: Routledge, 2003.

———. *Gesammelte Werke.* Edited by Beda Allemann and Stefan Reichert. Frankfurt am Main: Suhrkamp, 1983.

———. "The Meridian." In *Sovereignties in Question: The Poetics of Paul Celan*, by Jacques Derrida, edited by Thomas Dutoit and Outi Pasanen, 173–85. New York: Fordham University Press, 2005.

———. *Poems of Paul Celan.* Translated by Michael Hamburger. New York: Persea Books, 1988.

Foucault, Michel. *The Order of Things: An Archaeology of the Human Sciences.* New York: Random House, 1970.

Greisch, Jean. *Ontologie et temporalité. Esquisse d'une interprétation intégrale de* Sein und Zeit. Paris: PUF, 1994.

Gueroult, Martial. *Descartes' Philosophy Interpreted According to the Order of Reasons.* Translated by Roger Ariew. Minneapolis: University of Minnesota Press, 1984.

Hazard, Paul. *The European Mind, 1680–1715: The Critical Years.* Translated by J. Lewis May. New York: Fordham University Press, 1990.

Hegel, Georg Wilhelm Friedrich. *Phenomenology of Spirit.* Translated by A. V. Miller. New York: Oxford University Press, 1977.

Heidegger, Martin. *Being and Time.* Translated by Joan Stambaugh. Albany: SUNY Press, 1996.

———. "Lettre à Monsieur Beaufret." In *Lettre sur l'humanisme*, translated by R. Munier, 182–84. Paris: Aubier, 1964.

Henrich, Dieter. *Der ontologische Gottesbeweis.* Tübingen: Mohr, 1960.

Husserl, Edmund. *The Phenomenology of Internal Time-Consciousness.* Translated by James Churchill. Bloomington: Indiana University Press, 1966.

Janet, Pierre. *L'évolution de la mémoire et de la notion du temp.* Paris: Chahine, 1928.

Lenz, Siegfried. *The German Lesson.* Translated by Ernst Kaiser and Eithne Wilkins. New York: Hill and Wang, 1971.

Levinas, Emmanuel. *Proper Names.* Translated by Michael B. Smith. Stanford, CA: Stanford University Press, 1996.

Löwith, Karl. *Gott, Mensch und Welt in der Metaphysik von Descartes bis zu Nietzsche.* Göttingen: Vandenhoeck und Ruprecht, 1967.

Manzoni, Alessandro. *The Betrothed.* Translated by Bruce Penman. New York: Penguin, 1984.

Mathers, E. P., and J. C. Mardrus, eds. *The Book of Thousand Nights and One Night.* New York: Routledge, 1990.

Menninghaus, Winfried. *Schwellekund. Walter Benjamins Passage des Mythos.* Frankfurt am Main: Suhrkamp, 1986.

Merleau-Ponty, Maurice. *Phenomenology of Perception.* Translated by Colin Smith. New York: Routledge, 1996.

Nancy, Jean-Luc. *The Experience of Freedom.* Translated by Peter Fenves. Stanford, CA: Stanford University Press, 1994.

Opitz, M., and E. Wizilsa, eds. *Benjamins Begriffe.* Frankfurt am Main: Suhrkamp, 2000.

Ovid. *Metamorphoses.* Translated by Frank J. Miller. Cambridge, MA: Harvard University Press, 1966.

Pareyson, Luigi. "Heidegger: la libertà e il nulla." In *Annuario filosofico* 5, 9–29. Milan: Mursia, 1989.

———. *Ontologia della libertà.* Turin: Einaudi, 1995.

———. *Verità e interpretazione.* Milan: Mursia, 1972.

Perone, Ugo. "Al limite del finito." In *Ermeneutiche della finitezza*, edited by Giovanni Ferretti, 207–16. Pisa-Rome: Istituti editoriali e poligrafici internazionali, 1998.

———. "Il racconto della filosofia." In *Annuario Filosofico* 14, 61–98. Milan: Mursia, 1998.

———. "La rivelazione e il pensiero filosofico del Novecento." In *Il prisma della rivelazione. Una nozione alla prova di religioni e saperi*, edited by Sergio Sorrentino, 85–100. Assisi: Cittadella, 2003.

———. *La verità del sentimento*. Naples: Guida, 2008.

———. *Le passioni del finito*. Bologna: Dehoniane, 1994.

———. "Limite e soglia." *Filosofia e teologia* 1 (2004): 7–15.

———. "L'intersoggetività del raccontare." In *Intersubjectivité et théologie philosophique*, edited by Marco Maria Olivetti, 411–29. Padua: Cedam, 2001.

———. *Modernità e memoria*. Turin: SEI, 1987.

———. *Nonostante il soggetto*. Turin: Rosenberg and Sellier, 1995.

———. *Storia e ontologia*. Rome: Studium, 1976.

Poma, Iolanda. *Eresie della fenomenologia*. Naples: ESI, 1996.

Potok, Chaim. *The Chosen*. New York: Simon and Schuster, 1967.

———. *The Promise*. New York: Alfred Knopf, 1969.

Ringleben, Joachim. *Erfahrung Gottes im Denken. Zu einer neuen Lesart des Anselmschen Argumentes (Proslogion 2–4)*. Göttingen: Vandenhoeck und Ruprecht, 2000.

Röd, Wolfgang. *Der Gott der reinen Vernunft. Die Auseinandersetzung um den ontologischen Gottesbeweis von Anselm bis Hegel*. Munich: Beck, 1992.

Rostagno, Sergio. *Karl Barth*. Brescia: Morcelliana, 2003.

Rovatti, Pier Aldo. *La posta in gioco. Heidegger, Husserl, il soggetto*. Milan: Bompiani, 1987.

Schelling, Friedrich Wilhelm Joseph. *Sämtliche Werke*, edited by M. Schröter. Munich: Beck and Oldenbourg, 1927–1943.

Scholem, Gershom. *Major Trends in Jewish Mysticism*. New York: Schocken Books, 1941.

Sozzi, Lionello. *Vivere nel presente. Un aspetto della visione del tempo nella cultura occidentale*. Bologna: Il Mulino, 2004,

Tomatis, Francesco. *L'argomento ontologico. L'esistenza di Dio da Anselmo a Schelling*. Rome: Città Nuova, 1997.

Index